Sandra Van Vlierberghe, Arn Mignon (Eds.)
Superabsorbent Polymers

Also of interest

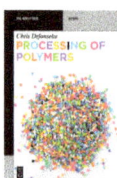

Processing of Polymers
Chris Defonseka, 2020
ISBN 978-3-11-065611-4, e-ISBN 978-3-11-065615-2

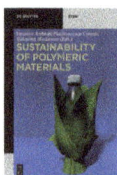

Sustainability of Polymeric Materials
Valentina Marturano, Veronica Ambrogi, Pierfrancesco Cerruti (Ed.),
2020
ISBN 978-3-11-059093-7, e-ISBN 978-3-11-059058-6

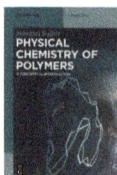

Physical Chemistry of Polymers.
A Conceptual Introduction
Sebastian Seiffert, 2020
ISBN 978-3-11-067280-0, e-ISBN 978-3-11-067281-7

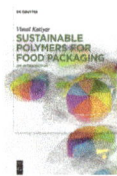

Sustainable Polymers for Food Packaging.
An Introduction
Vimal Katiyar, 2020
ISBN 978-3-11-064453-1, e-ISBN 978-3-11-064803-4

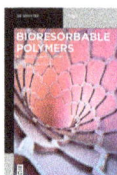

Bioresorbable Polymers.
Biomedical Applications
Declan Devine (Ed.), 2019
ISBN 978-3-11-064056-4, e-ISBN 978-3-11-059709-7

Superabsorbent Polymers

Chemical Design, Processing, and Applications

Edited by
Sandra Van Vlierberghe and Arn Mignon

DE GRUYTER

Editors

Prof. Dr. Sandra Van Vlierberghe
Polymer Chemistry & Biomaterials Group
Centre of Macromolecular Chemistry
Ghent University
Krijgslaan 281, S4-Bis
9000 Ghent
Belgium
http://pbmugent.eu
Sandra.VanVlierberghe@UGent.be

Prof. Dr. Ir. Arn Mignon
Smart Polymeric Biomaterials Research Group
Catholic University Leuven (KU Leuven)
Andreas Vesaliusstraat 13
3000 Leuven
Belgium
Arn.Mignon@KULeuven.be

ISBN 978-1-5015-1910-9
e-ISBN (PDF) 978-1-5015-1911-6
e-ISBN (EPUB) 978-1-5015-1171-4

Library of Congress Control Number: 2020948027

Bibliographic information published by the Deutsche Nationalbibliothek
The Deutsche Nationalbibliothek lists this publication in the Deutsche Nationalbibliografie;
detailed bibliographic data are available on the Internet at http://dnb.dnb.de.

© 2021 Walter de Gruyter GmbH, Berlin/Boston
Cover image: malik parwaiz akhter/iStock/Getty Images Plus
Typesetting: Integra Software Services Pvt. Ltd.
Printing and binding: CPI books GmbH, Leck

www.degruyter.com

Contents

List of contributing authors

María Fernanda Baieli
Facultad de Farmacia y Bioquímica
Universidad de Buenos Aires and Instituto de
Nanobiotecnología (NANOBIOTEC),
CONICET-UBA
Junín 956
(CC1113AADEGA) Buenos Aires
Argentina
fer.baieli@gmail.com

Filis Curti
The Advanced Polymer Materials Group
University POLITEHNICA of Bucharest
Bucharest, Romania

Marie-Christine Durrieu
Chimie et Biologie des Membranes et Nano-
Objets (UMR5248 CBMN)
Université de Bordeaux
Allée Geoffroy Saint Hilaire - Bât B14
33600 Pessac
France

Gaétan Laroche
Laboratoire d'Ingénierie de Surface
Centre de Recherche sur les Matériaux
Avancés
Département de Génie des Mines de
la Métallurgie et des Matériaux
Université Laval
1065 Avenue de la médecine
Québec G1V 0A6
Canada

Carmen Nicolae
The Faculty of Medical Engineering
University POLITEHNICA of Bucharest
Bucharest, Romania

Elena Olăret
The Advanced Polymer Materials Group
University POLITEHNICA of Bucharest
Bucharest, Romania

Emilie Prouvé
Laboratoire d'Ingénierie de Surface
Centre de Recherche sur les Matériaux
Avancés
Département de Génie des Mines de
la Métallurgie et des Matériaux
Université Laval
1065 Avenue de la médecine
Québec G1V 0A6
Canada

Christof Schröfl
Institute of Construction Materials
Faculty of Civil Engineering
Technische Universität Dresden
Georg-Schumann-Str. 7
DE-01187 Dresden
Germany
Christof.Schroefl@tu-dresden.de

Andrada Serafim
The Advanced Polymer Materials Group
University POLITEHNICA of Bucharest
Bucharest, Romania

Didier Snoeck
Magnel-Vandepitte Laboratory
Department of Structural Engineering and
Building Materials
Faculty of Engineering and Architecture
Ghent University
Technologiepark Zwijnaarde 60
B-9052 Ghent
Belgium
didier.snoeck@ugent.be

https://doi.org/10.1515/9781501519116-203

Izabela-Cristina Stancu
The Faculty of Medical Engineering
University POLITEHNICA of Bucharest
Bucharest, Romania

Nicolás Urtasun
Universidad de Buenos Aires,
Facultad de Ciencias Exactas y Naturales and
Instituto de Química Biológica de la Facultad
de Ciencias Exactas y Naturales (IQUIBICEN),
CONICET-UBA
Intendente Guiraldes 2160
(C1428EGA) Buenos Aires
Argentina
nurtasun@conicet.gov.ar

Christof Schröfl

1 Chemical design and synthesis of superabsorbent polymers

Abstract: Chemical design and realization of superabsorbent polymers (SAPs) are driven by most appropriate performance of the products in distinct applications. Speed and final level of absorption as well as retention efficiency depend on the ionic composition of the solutions to which the polymer is typically exposed. Requirements originating from applications in sanitary products and healthcare have dominated ever since. Further target fields include agriculture and cement-based building materials.

Superabsorbers that are available are mostly based on acrylate chemistry, and these are synthesized through free-radical copolymerization in bulk/solution or inverse suspension processes. Comonomers can be integrated into feature-specific effects, for instance, sulfonic or phosphonic, in addition to or instead of carboxylate groups, cationic sites to form amphoteric polymers, nonionic but hydrophilic moieties to dilute the density of the ionic density along the polymer chains, or biologically active motives. The primary products are commonly processed further by elaborate postpolymerization treatments. Most important is surface postcross-linking to minimize the stickiness of powdery superabsorbent products and the so-called gel-blocking effect in superabsorber beds.

Research has recently focused on biopolymers and bioderived main monomers and cross-linkers. Polysaccharides, such as cellulose, starch, alginate, chitin, and chitosan, play most important roles, and proteins have also been utilized. These natural substances can chemically be slightly modified or grafted by acrylate-based network structures. A broad range from distinctly biobased superabsorbers to hybrids from petrochemical and biogenic sources can be generated.

1.1 Thermodynamics and kinetics of liquid absorption by superabsorbent polymers

SAP particles take up aqueous solution in considerably larger mass and volume than their individual dry weight. The absorption kinetics, final amount absorbed, and the stability of the swollen product, that is, the hydrogel, depend on molecular composition and grading as well as the ionic loading of the liquid phase. The polymer particles remain individually particulate in their water-enriched state, because the cross-linked primary polymer chains represent a three-dimensional network

https://doi.org/10.1515/9781501519116-001

which is insoluble as an entity. Thermodynamically, Equation (1.1) describes the equilibrium state [1]:

$$\Delta\pi = \Delta\pi(\text{mix}) + \Delta\pi(\text{elastic}) + \Delta\pi(\text{ion}) + \Delta\pi(\text{bath}) = 0 \qquad (1.1)$$

At first, polymer–solvent interactions induce osmotic pressure $\Delta\pi(\text{mix})$, which makes the network to expand. Solvent molecules attach to each primary chain and, potentially, cross-linking units of the SAP. However, these chains cannot move freely – which would be denominated as dissolving – due to their chemically stable interconnections. A counteracting elastic force $\Delta\pi(\text{elastic})$ builds up, which limits the overall expansion. Next, ionic groups along the chains deprotonate or protonate in nonneutral aqueous solution. This makes the chains charged likewise, and they repel each other. Another expansive pressure, $\Delta\pi(\text{ion})$, initiates. The charge-balancing, dissolved ions are mobile but they cannot leave the interior of the expanded network without replacement by other ions. The hydrogel itself must remain electronically neutral as an entity. Ingressing ions from the solution outside can penetrate the hydrogel, which results in polyelectrolyte shielding $\Delta\pi(\text{bath})$. The osmotic pressure is reduced pronouncedly and may even be contractive [1]. Taking into account these stepwise processes, an optimum swelling speed, final level of absorption, and, potentially, even delayed self-releasing properties can be adjusted via chemical composition and grading of an SAP product in the context of its intended application and conditions of its surroundings.

Besides diligently designing one SAP species for most beneficial sorption behavior in a distinct field of application, the sorption properties of an SAP product can be fine-tuned by blending two or more well-defined types of SAP [2].

1.2 Polymer synthesis of petrochemical-derived SAPs

1.2.1 Radical (co)polymerization

SAPs have prevailingly been synthesized via free-radical (co)polymerization, the reaction mechanism being well understood and mathematically modeled [3]. In brief, the fundamental steps are:
- Initiation: Cleavage of an initiator molecule forming a radical, followed by bonding of this initiator radical to one monomer molecule, forming a radical site at the monomer.
- Chain growth by continued addition of monomer molecules forming the trunk chains, whereby comonomers integrate; multifunctional comonomers may integrate in the primary chains and offer reactive sites for branching at their pending functionality, that is, act as a cross-linking point.
- Termination by reaction of two radical sites with each other or by disproportionation.

- Postpolymerization steps like cleaning from residual monomers and further chemical processing with other substances by different types of reaction mechanisms.

Using this reaction pathway, there are two prominent modes for synthesis of SAP from procedural point of view:

- Block (co)polymerization, or (co)polymerization in aqueous solution, resulting in a gel block, followed by drying and disintegrating the macroscopic, three-dimensional body by milling, grinding, and sieving; irregular particles are received and particle size distributions are adjusted in the postsynthesis processing steps [3–12].
- Inverse suspension or emulsion (co)polymerization to prepare spherical particles; the particle size distribution and microstructure are determined by synthesis parameters (e.g., agitation intensity, streaming conditions), choice of (co)solvents, as well as chemical additions (e.g., types and dosages of emulsifiers) [3, 13–18].

Postpolymerization treatments include reactive options, for example, surface post-cross-linking, and reprocessing steps like drying to a distinct moisture content.

Looking backward from the utilization point of view, superabsorbers are optimized toward characteristic fields of application, each of which may feature peculiarities in terms of ionic composition and pH of the liquid to be uptaken, temperature, pressure, and pressure gradients. In medical and hygiene or healthcare products, in which quick and safe absorption and storage of any kind of body fluids from human beings at most specific temperatures are essential, tolerance by the skin is also an important factor. The latter requirement is less important for SAP in other large-scale applications such as agriculture and horticulture, food processing, fresh and wastewater treatment, or civil engineering. However, each of these fields may impose other typical requirements on the SAP [3, 11, 19–48].

In free-radical copolymerization, the reaction rate is uncontrolled and fairly high, resulting in a not well-controlled polymer structure. Miyajima et al. described the application of iodine transfer polymerization for the synthesis of acrylate-based SAP. The progress of the polymerization is well controlled by implementing an organoiodine compound as a chain transfer agent. Such compounds need to exhibit groups that stabilize free-radical chain ends to increase the transfer, especially at the early stage of polymerization. Figure 1.1 shows exemplary partial structural formulae, as disclosed in the respective publication, which does not name specific substances but only characteristic motives. Solely, 3-iododihydrofuran-2(3H)-one has been sketched as a distinct molecule. Furthermore, they must be soluble in water and stable against hydrolysis. Apart from a more homogeneous polymer structure, the SAP products possessed pronouncedly increased absorption characteristics. This method was selected among other reversible-deactivation radical

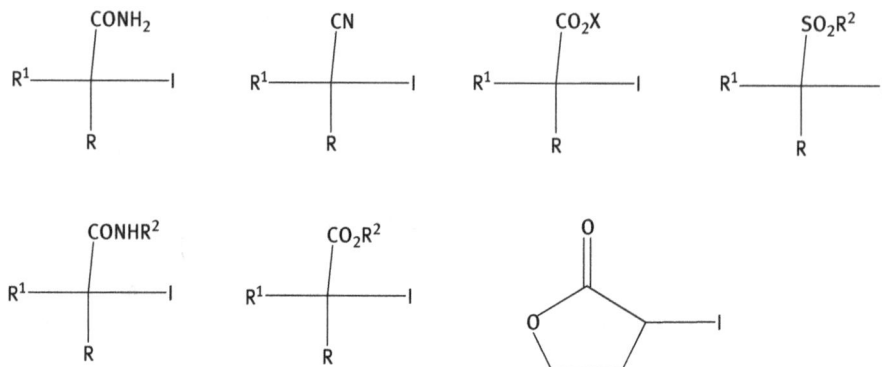

Figure 1.1: Organoiodine species for iodine transfer polymerization of acrylic acid-based SAP as summarized in [49].

polymerization strategies for practical reasons like no discoloration, nontoxicity, cost, and no unnecessary additional purification steps [49].

1.2.2 Initiators, comonomers and cross-linkers in acrylate-based SAP

Patent literature originating in the large-scale industrial sector majorly focuses on SAP based on acrylic acid as the main monomer. Academic publications naturally describe educts, procedures, processes, and additives more distinctly. Fundamental substances and synthesis techniques include both industry and academia [3, 6, 7, 10–13, 18, 23–42, 45, 47, 50–55]:

The (co)polymerization is **initiated** by
- Free-radical initiators: Peroxodisulfate, *tert*-butyl hydroperoxide, hydrogen peroxide, and 2,2-azobis(2-amidinopropane)dihydrochloride
- Redox systems are composed of a free-radical initiator and a reducing chemical such as sodium sulfite, sodium hydrogensulfite, (2R)-2-[(1S)-1,2-dihydroxyethyl]-3,4-dihydroxy-2H-furan-5-one (i.e., L-ascorbic acid), some reducible ferrous salt, or various amines
- Irradiation with high-energy rays like radiations, electron beam, or ultraviolet (UV) rays (wavelength 100–400 nm)

Main monomer: Acrylic acid (propenoic acid), methacrylic acid (2-methylprop-2-enoic acid), or their alkali salts, which may beneficially contain a well-defined amount of inhibitor such as *ortho-/meta-/para*-methoxyphenol, 5-methylidenefuran-2-one (i.e., protoanemonin), or furan-2-carbaldehyde (i.e., furfural). Acrylates and methacrylates may be used, which are indicated by { } brackets in the text.

Besides fossil substances, acrylic acid can be prepared from nonfossil raw materials. However, such sources may contain fairly high contents of propanoic (propionic) acid, which is unfavorable for radical polymerization reactions. Propanoic acid is a by-product in the preparation of acrylic acid from acetic acid, propane-1,2,3-triol (i.e., glycerine, which by itself stems from fats and oils in biodiesel production), or 3-hydroxypropanoic acid (generated from cellulose or glucose fermentation). Up to 0.5 wt% (percent by weight) of propanoic acid with respect to acrylic acid has been reported to be tolerable. On the other hand, a remainder of 10–100 ppm (parts per million) of propanoic acid in the final SAP produce is tolerated because it causes antibacterial properties, whereas malodor or miscoloring does not occur at such low concentrations.

Comonomers are incorporated to dilute the density of the carboxylic groups along the polymeric backbone and to introduce further types of chemical functional groups:
- Anionic: Hydrophilic C=C unsaturated sulfonic acids such as vinyl sulfonic acid, styrene sulfonic acid, and 2-acrylamido-2-methylpropanesulfonic acid (AMPS)
- Cationic: Hydrophilic quaternary amines with one C=C unsaturated moiety
- Nonionic but hydrophilic: Substituted acrylamides, hydroxyalkyl-{meth}acrylates, {ω-methoxy} polyethylene glycol {meth}acrylate, vinyl pyridine, N-vinylpyrrolidone, N-acryloylpiperidine, N-acryloylpyrrolidine, and N-vinylacetamide

Cross-linking forms the superabsorbent primary product. Internal cross-linkers possess two or more polymerizable C=C unsaturated, or otherwise reactive, moieties to react with two or more primary chains and form a chemically stable three-dimensional network:
- Di-, tri-, or oligo-{meth}acrylates or {meth}acrylamides, mostly, N,N-methylenebisacrylamide (MBA), but also for example, 2-[2-(2-prop-2-enoyloxyethoxy)ethoxy]ethyl prop-2-enoate (i.e., triethylene glycol diacrylate)
- Di- or triallyl compounds such as 2,4,6-tris(prop-2-enoxy)-1,3,5-triazine (i.e., triallyl cyanurate), 1,3,5-tris(prop-2-enyl)-1,3,5-triazinane-2,4,6-trione (i.e., triallyl isocyanurate), 1,3,5-tris(2-methylprop-2-enyl)-1,3,5-triazinane-2,4,6-trione (i.e., trimethylallyl isocyanurate), triallylphosphate, triallylamine, and poly{meth}allyloxyalkanes
- Substituted diglycidyl ethers or glycols or glycerols
- 2,2-Bis(hydroxymethyl)propane-1,3-diol (i.e., pentaerythritol), or ethylene diamine, or any alkylene carbonates

1.2.3 Use of other C=C unsaturated carboxylic acids than acrylic acid

Besides monocarboxylic acid, C=C unsaturated bicarboxylic acids can be applied as the main monomer in the synthesis strategy outlined above. Comonomers and cross-linkers can be varied at the same time [3]. 2-Methylidenebutanedioic acid (i.e., itaconic acid) is a promising main monomer; the major advantage of it being its straightforward biosynthesis by bacteria and good biodegradability of polymers made of it [54]. Any comonomers denominated above can be used. Exemplarily, acrylamide-co-itaconic acid cross-linked with MBA has been described as a well-working SAP [56].

1.2.4 SAP with sulfonic moieties instead of carboxylic groups

For use in cement-based building composites, superabsorbers with sulfonic instead of carboxylic moieties have been disclosed. Their sorption properties are less sensitive to the Ca^{2+} dissolved in the liquid phase of such paste- or suspension-like materials because the sulfonic group cannot be complexed with Ca^{2+}. AMPS is used as a typical main monomer for such SAPs instead of the acrylic acid [57] or as a comonomer to acrylic acid [55]. The fundamental reactions and chemicals such as initiators, comonomers, or cross-linkers of the AMPS-based SAP are in accord with the acrylate-based ones.

A powdery accelerator mixture for cement-based building materials containing nanosized calcium silicate hydrate phases (C–S–H) immobilized in SAP granules has been described by Langlotz et al. [43]. It is specifically bound for dry mix mortar preblends. Synthetic C–S–H particles (diameter smaller than 200 nm) are dispersed in an aqueous solution and then immersed in the swellable SAP particles. After drying to a residual water content of approximately 10 wt% at temperatures below 80 °C, a powdery product is obtained. Presumably, the C–S–H particles migrate out of the hydrogel after contact with the mixing water of the cement-based composite and act as nucleation sites for hydration products. Specific sulfonate monomers can be AMPS, 2-methacrylamido-2-methylpropanesulfonate, 2-acrylamidobutanesulfonate, or 2-acrylamido-2,4,4-trimethylpentanesulfonate.

An SAP with distinctly delayed swelling for oil well cementing has been designed with AMPS. The most important monomers are AMPS, MBA, acrylate, and two kinds of cross-linkers. The first cross-linking monomer possesses ester bonds that are cleavable when in contact with cement pore solution or brine in time. The typically elevated temperatures in boreholes of up to far more than 120 °C support this reaction and a delayed swellability of the SAP results. However, such intensified swelling has to start during or after setting of the cement but must not start too early. Potential chemicals include di-, tri-, or tetraacrylates (e.g., ethylene diacrylate), polyethylene glycol di{meth}acrylate (number of ethylene glycol units being 2–30), glycerol dimethacrylate (GDMA), triglycerol diacrylate, ethoxylated glycerol triacrylate, {ethoxylated} pentaerythritol tetraacrylate, pentaerythritol triacrylate, and {ethoxylated} trimethylolpropane triacrylate. The second cross-linker is a compound that forms chemically stable bonds between the primary chains, specifically,

N,N'-methylenebis-{meth}acrylamide, *N,N'*-(1,2-dihydroxy-1,2-ethanediyl)bisacryla-mide, *N,N'*-(1,2-ethanediyl)bisacrylamide, *N,N'*-[[2,2-bis(hydroxymethyl)-1,3-pro-panediyl]bis(oxymethylene)]bisacrylamide, bis-(2-methacryloyl)oxyethyl disulfide, *N,N'*-bis(acryloyl)cystamine, or 1-ethenylsulfonylethene (i.e., divinyl sulfone) [44].

Besides application in building materials, SAPs with AMPS have been described efficient in wastewater treatment, that is, removal of inorganic and organic solutes from aqueous solution [47].

1.2.5 SAP with phosphonic moieties instead of carboxylic groups

Besides carboxylate and sulfonate, the phosphonic group has been applied as the major anionic functionality in SAP. 1-Hydroxyethylidene-1,1'-diphosphonic acid has been claimed preferable by Hartnagel et al. [58]. Advantages of such SAP improved the stability of color, that is, avoiding miscoloring, and enhanced permeability of the gel during its formation in the synthesis and further processing steps [58].

1.2.6 Incorporation of purposeful comonomers for distinct functionalities

SAP samples with fine-tuned viscoelastic characteristics were prepared from acrylam-ide and *N*-(4-ethylphenyl)acrylamide, which represents a hydrophobic substance and is incorporated in the backbones in small amount, and a C=C-unsaturated acid as a further essential comonomer (Figure 1.2). The acidic comonomers were either acrylic acid, AMPS, or 4-ethenylbenzenesulfonic acid (i.e., 4-styrenesulfonic acid) [59].

Poly(acrylamide-*co*-sodium 4-hydroxy-2-methylenebutanoate) hydrogel show en-hanced phytotoxicological efficiency in a study bound to elucidate most suitable SAPs for plant growth substrates with optimized water storage capacity for this field of application. This polymer performed best when compared with polyvinyl alcohol (PVA), poly(acrylamide), and poly(acrylamide-*co*-sodium acrylate), the latter three being cross-linked with borate [46].

An antimicrobial SAP was synthesized via cross-linking of a prepolymer prepared from *trans*-1,4-dibromo-2-butene and *N,N,N',N'*-tetramethyl-1,3-propanediamine with tris(2-aminoethyl)amine. This cationic network was then swollen by 2-hydroxyethyl acrylate and [2-hydroxy-3-(2-methylprop-2-enoyloxy)propyl] 2-methylprop-2-enoate (i.e., GDMA) via photopolymerization and an interpenetrating hydrogel was formed (Figure 1.3) [60].

Pronouncedly pH-dependent swelling is very important for self-sealing and self-healing of cracks in cement-based building materials. The collapsed state should not swell in the high alkaline pH during mixing but only later on do so when rather neutral water enters the structural member. Exemplarily, SAP prepared

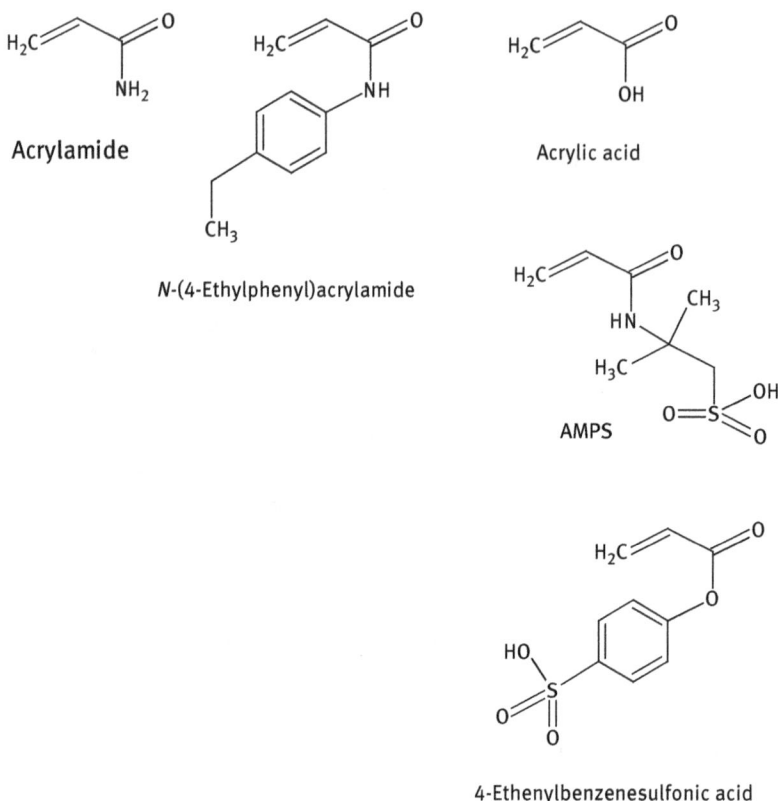

Acrylamide

N-(4-Ethylphenyl)acrylamide

Acrylic acid

AMPS

4-Ethenylbenzenesulfonic acid

Figure 1.2: Monomers for SAP synthesis: nonionic but hydrophilic acrylamide, hydrophobic N-(4-ethylphenyl)acrylamide, and one acidic component such as acrylic acid, AMPS, and/or 4-ethenylbenzenesulfonic acid as extracted from ref. [59].

from the comonomers, 2-(dimethylamino)ethyl 2-methylprop-2-enoate and trimethyl(2-prop-2-enoyloxyethyl)azanium chloride, implementing the cross-linker MBA, initiated by ammonium peroxodisulfate, proved especially efficient for this application [48].

1.2.7 Nonionic synthetic SAP based on poly(ethylene glycol) diacrylate

Aiming at biomedical produces, poly(ethylene glycol) has been proposed as the fundamental primary chains for the network formation. Diacrylated poly(ethylene glycol) with a molecular weight of about 700 g/mol was physically polymerized as well as cross-linked by UV [61].

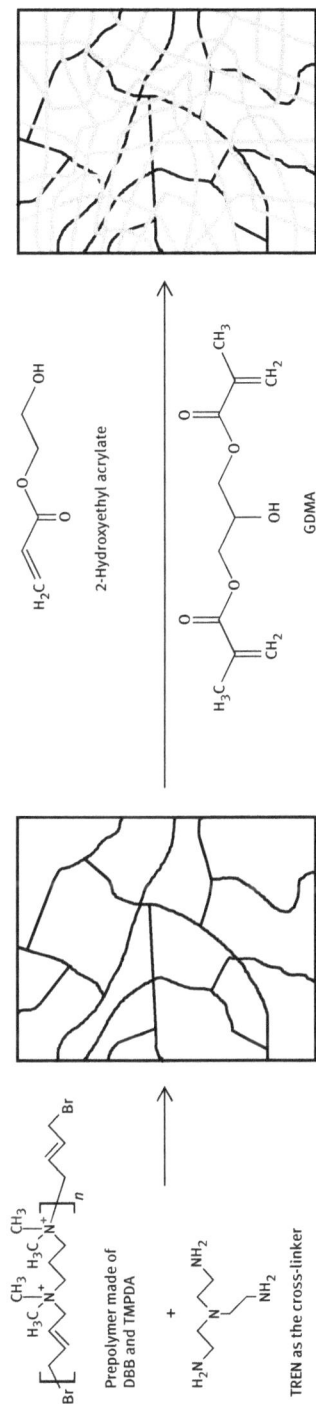

Figure 1.3: Two-step procedure to generate an antimicrobial interpenetrating network hydrogel as described by Strassburg et al. [60].

1.2.8 Postpolymerization treatments

1.2.8.1 Conventional strategies and chemicals

The primary SAP can be used as obtained. However, postpolymerization treatments are commonly applied to generate practically applicable particles, improve distinct features, or merely avoid problematic properties [6, 7, 13, 18, 23–42, 45, 50–53].

With gel-polymerized polymer blocks, such procedures typically include:
- Removal of remaining unreacted or potentially released liquid substances
- Crushing, grinding, milling, and sieving to achieve desired gradings
- Drying at 120–200 °C to achieve well-defined moisture contents and fine-tune physical properties for further processing, which is a first drying step that can be complemented by further drying right at the end of all processing just before delivery.

With (inverse) suspension-polymerized primary SAPs, crushing, milling, and sieving are not required because the synthesis reveals individual particles of intended size. However, the other steps of chemical cleaning and adjustment of moisture content apply for them as well.

Reactive, adsorptive, or blending steps may follow to add further functionalities or adjust specific properties:

Surface postcross-linking, which typically affects about 10% of the particle diameter, pronouncedly reduces the so-called gel-blocking effect. The "gel-blocking" effect describes the phenomenon of limited absorbency of a collective of primary SAP particles because the liquid is too quickly and intensely absorbed very close at its entry point. These saturated particles tend to efficiently inhibit the liquid penetration deeper inside the probe and hence, the entire SAP material is not available for absorption. The individual SAP particles must be modified to let pass a certain amount of liquid downstream. Surface postcross-linking is an efficient measure to enhance the permeability of a macroscopic batch of individual SAP particles. Stickiness and permeability of the particulate intermediate and final products are improved as well. Strategies described include chemical reaction at an elevated temperature and/or at a reduced pressure. Catalysts may be present as well. Exemplary reactants are
- Hydrophilic compounds featuring at least two hydroxyl groups: (substituted) diols, triols, or polyols, such as ethane-1,2-diol, propane-1,3-diol, propane-1,2,3-triol, and butane-1,4-diol, as well as di- or triethanol amine
- Epoxy compounds with ester or ether moieties, for example, in form of glycidyl esters
- Some polyamine or its inorganic or organic salt
- A polyisocyanate like 2,4-tolylene diisocyanate or hexamethylene diisocyanate
- Some polyoxazoline, for example, 1,2-ethylenebisoxazoline

- An alkylene carbonate
- Some halo-epoxy compounds such as 2-(chloromethyl)oxirane (i.e., epichloro-hydrin), 2-(chloromethyl)-2-methyloxirane (i.e., α-methylepichlorohydrin), 2-(bromomethyl)oxirane (i.e., epibromohydrin), or respective polyamine adducts
- A silane, for example, γ-glycidoxypropyltrimethoxysilane or γ-aminopropyl triethoxysilane;
- Inorganic water-soluble hydroxide or chloride salt with a multivalent metal cation (Mg^{2+}, Al^{3+}, Ca^{2+}, Ti^{4+}, Fe^{2+}, Fe^{3+}, Zn^{2+}, or Zr^{4+}), by name, mostly cited, aluminum sulfate
- X-ray amorphous $Al(OH)_3$
- For irradiation with UV (100–400 nm) in conjunction with a polyunsaturated organic compound, potentially supplemented by a Brønsted or a Lewis acid (aluminum sulfate) as a catalyst: allyl-based substances including diallyl dimethyl ammonium chloride, triallyl cyanurate or isocyanurate, pentaerythritol-triallyl- or tetraallylether, tetraallylorthosilicate, tetraallyloxyethane, diallylphthalate, triallylamine, and triallylcitrate; various hydrophilic di-, tri-, or oligoacrylate esters; allyl methacrylate, a squalene, MBA, eicosapentaenoic acid, (2E,4E)-hexa-2,4-dienoic acid (i.e., sorbic acid), a vinyl-terminated silicone, or a polysiloxane carrying a quaternary ammonium group
- For irradiation with microwaves, using N,N-dimethylaniline as a catalyst: 2-[4-(oxiran-2-ylmethoxy)butoxymethyl]oxirane (i.e., 1,4-butanediol diglycidyl ether), 2-[2-(oxiran-2-ylmethoxy)ethoxymethyl]oxirane (i.e., ethyleneglycol diglycidyl ether)

Another prominent surface postsynthesis treatment is the addition of 10–990 ppm of water-insoluble **inorganic (nano)particles**. They will reside on or close to the surface of the SAP particles and mitigate the so-called gel-blocking effect. SiO_2, with a specific surface area between 30 and 330 m^2/g (square meters per gram, measured according to the Brunauer–Emmett–Teller method), potentially having quaternary amino groups on their particle surface to create ionic bonds to the polymer resin, has frequently been focused on as well. TiO_2 or any comparable nanoparticles have been described or claimed for this step in the creation process of SAP products.

Besides, a **surfactant** to hydrophobize the outer surface of the particles to a well-balanced extent can be bonded to the SAP surface. This measure as well mitigates the so-called gel-blocking effect. Such surfactant contains a functional group to bind to the SAP resin. An ionic bond can be formed at reaction temperatures below 100 °C, accomplished by primary, secondary, or tertiary amines. Alternatively, the surfactant binds covalently at a reaction temperature between 120 and 240 °C, using a carboxylic diester or a diol. The hydrophobic part of the surfactant is typically an aliphatic hydrocarbon C_{16}–C_{24} or poly(ethylene oxide)-co-poly(propylene oxide).

Next, a **cationic polymer compound** can be incorporated in the surface region of the primary SAP particles to increase the retention capacity and simultaneously suppress the "gel-blocking" effect. Potential educts have been described as
- Polyalkyleneimines
- Various polyamines, such as polyether polyamines, polyvinylamines, polyalkyl-amines, polyallylamines, polydiallylamine, poly(N-alkyl allylamine), monoallyl-amine-diallylamine copolymer, N-alkylallylamine-monoallylamine copolymer, monoallylamine-dialkyldiallyl ammonium salt copolymer, diallylamine-dialkyl-diallyl ammonium salt copolymer, and polyalkylene polyamines
- Polyamide amine grafted with ethylene imine or a protonated polyamide amine
- Condensates of polyamide or amine with epichlorohydrin
- Distinctly, a salt of poly(vinylbenzyldialkylammonium) or poly(2-hydroxy-3-methacryloyloxypropyl-dialkylamine), as well as partially hydrolyzed poly(N-vinylformamide), partially hydrolyzed poly(N-vinylalkylamine), partially hydrolyzed copolymer of N-vinylformamide and N-vinylalkylamide
- Polyvinyl-imidazole, -pyridine, -imidazoline, -tetrahydropyridine; polydialkyla-mino-alkylvinylether, polydialkylaminoalkyl{meth}acrylate, polyallylamine, and polyamidine

Preferred hydrophilic coupling agents include epoxides (i.e., alkylene oxides, styrene oxide, and 1-phenylpropylene oxide), various glycidyl ethers, or silanes with epoxide groups.

A **combination of a multivalent cation and an organic acid** (favorably hydrocarbon C_9-C_{29}) has also been described to increase the retention capacity and simultaneously suppress the "gel-blocking" effect. The surface of an already post-cross-linked SAP is reacted at 40–100 °C for 10–120 min with:
- Aluminum sulfate or a cationic polymer with a solubility of less than 1 g per 100 g water at 25 °C, combined with a polyethylenimine, polyalkylamine, modified polyamideamine denaturalized by graft of ethylenimine, protonated poly-amideamine, condensates of amines or polyamideamine and 2-(chloromethyl) oxirane epichlorohydrin, poly(vinylbenzyldialkylammonium) salt, poly(diallyl-alkylammonium) salt, poly(2-hydroxy-3-methacryloyloxypropyldialkyl amine), polyetheramine, modified polyvinylamine, partial hydrolysate of poly(N-vinyl-formamide), partial hydrolysate of poly(N-vinylalkylamide), partial hydrolysate of a copolymer of (N-vinylformamide) and (N-vinylalkylamide), polyvinylimid-azole, polyvinylpyridine, polyvinylimidazoline, polyvinyltetrahydropyridine, polydialkyl-aminoalkylvinylether, polydialkylaminoalkyl{meth}acrylate, poly-allylamine, or polyamizine
- Cationic modified starch or cellulose; and an organic acid
- One of the fatty acids
- A petroleum acid, for example, benzoic acid, naphthoic acid, and naphthoxy-acetic acid

- A noncarboxylic organic acid, for example, alkylsulfuric acid (C_9 or longer), alkylbenzene sulfonic acid, alkylphosphonic acid, alkylphosphine acid, or alkylphosphoric acid

Further postsynthesis treatments can be accomplished to improve other properties or to avoid chemical or physical drawbacks. Hereby, the SAPs that have already passed the procedures outlined so far now react with:
- Chelating agents to limit the free contents of Fe^{2+}, Fe^{3+}, and Cr^{3+} or other metal ions that are not intentionally used for surface postcross-linking. Such metal ions may cause degradation of primary polymer chains and give rise to anesthetic discoloring of the final product. Chemicals include 2,2',2'',2'''-(ethane-1,2-diyldinitrilo)tetraacetic acid (EDTA), [2-[bis(phosphonomethyl)amino] ethyl-(phosphonomethyl)amino]methylphosphonic acid, or a mixture of (1-hydroxy-1-phosphonoethyl)phosphonic acid (i.e., etidronic acid) and 2-hydroxy-2-sulfonatoacetic acid or their respective salts
- Biocides and antibacterial admixtures to suppress decomposition or rotting, for example, hexanoic acid, butanoate, or 3-methylcyclopent-2-en-1-one
- Deodorant substances like extracts of bamboo or tea

At the very end of the production process, SAPs are readjusted to a certain moisture content. This, on the one hand, avoids dust formation and, on the other hand, equalizes the absorption speed and quantity all over the bulk volume of particles [8, 62].

Few distinct and recent examples of postpolymerization treatments following conventional physical or chemical approaches should finally be mentioned by name:
- Magnetic fields applied to extract metallic impurities from the product [9]
- Intercalation into graphite [17]
- Implementation of lithium chloride to boost moisture absorption and use of the product as an air-desiccating agent [63]
- Incorporation of Fe_3O_4 particles to form magnetic SAP for use as easily separable water extracting agent from oil–water mixtures [64]

1.2.8.2 "Click" reaction as a reactive postpolymerization treatment

Moini et al. synthesized acrylate-based SAP cross-linked with poly(ethylene glycol)-dimethacrylate and applied a catalytic "click" reaction for surface postcross-linking [65]. The latter was a two-step process of a solvent-induced phase transition followed by copper-catalyzed azide–alkyne cycloaddition. 1,2,3-Triazole groups were formed in the surface region of the SAP particles (Figure 1.4). The product showed

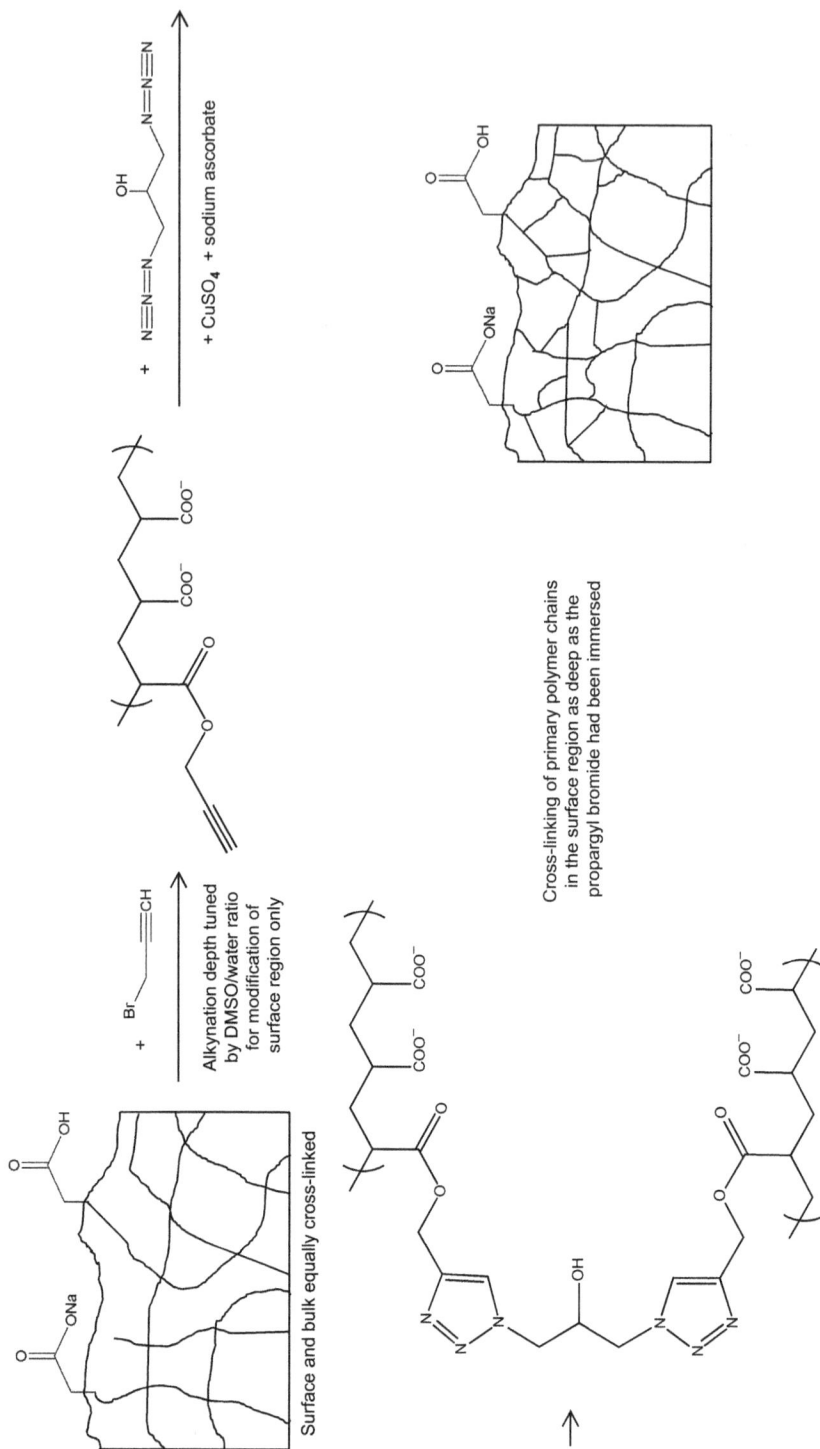

Figure 1.4: Reaction scheme of the "click" reaction postpolymerization treatment to introduce 1,2,3-triazole groups in the surface region of acrylate-based SAP particles as outlined by Moini et al. [65].

improved mechanical properties, that is, increased storage modulus of the swollen polymer as assessed by oscillatory rheometry, and highly intensified antibacterial activity [65].

1.2.9 Composites of SAP with inorganic and organic particles

A filler material can be integrated inside the SAP network structure to improve the absorption capacity and increase the mechanical strength.

SAPs were wrapped around TiO_2 nanoparticles. The SAP gel was based on acrylic acid and AMPS, and 3-triethoxysilylpropyl-2-methylprop-2-enoate acted as a coupling agent. Motivation for this core–shell structure was to incorporate it into an epoxy resin as a self-healing promoter [66].

Some natural pozzolan was incorporated as an inorganic filler into acrylic acid-co-acrylamide-based SAP during the main synthesis of solution/bulk polymerization. Interestingly, this study used a special type of cross-linker, by name 2-[hydroxy-[2-(2-methylprop-2-enoyloxy)ethoxy]phosphoryl]oxyethyl 2-methylprop-2-enoate (i.e., bis (methacryloyloxyethyl) hydrogen phosphate) [10].

SiO_2 as the inorganic component was used both for the core and the shell in composite SAPs. To generate core–shell particles with a SiO_2 core covalently bonded inside a superabsorbent shell, silica particles were prepared in a sol–gel process from tetraethyl orthosilicate (TEOS) and vinyl triethoxysilane first. The silica particles thus featured vinyl groups. In the second step, these groups were polymerized with acrylamide and MBA as the cross-linker in a free-radical reaction initiated by potassium peroxodisulfate (Figure 1.5) [67].

In an opposite approach, SAP was based on poly(methacrylate-co-ethylene glycol dimethacrylate) used as a template, or core, in the synthesis of SiO_2 nanoparticles. The sol–gel process of SiO_2 formation from TEOS was modified by placing the SAP in the solution and, hence, to generate the SiO_2 attached to the polymer granules. Such produce can prospectively be utilized as a self-healing promoter in cement-based building materials [68].

As a biobased support inside SAP, sawdust from fir was used. It was simply present as well in the monomer mixture when synthesizing the SAP from acrylic acid, acrylamide, and MBA by radical copolymerization in solution [69].

1.2.10 Polyvinyl alcohol-based SAP

PVA substituted with styrylpyridinium groups through two ether bonds was cross-linked to an SAP material by irradiation with UV. Due to insufficient mechanical strength, cellulose nanocrystals were incorporated and composite hydrogel was formed [70].

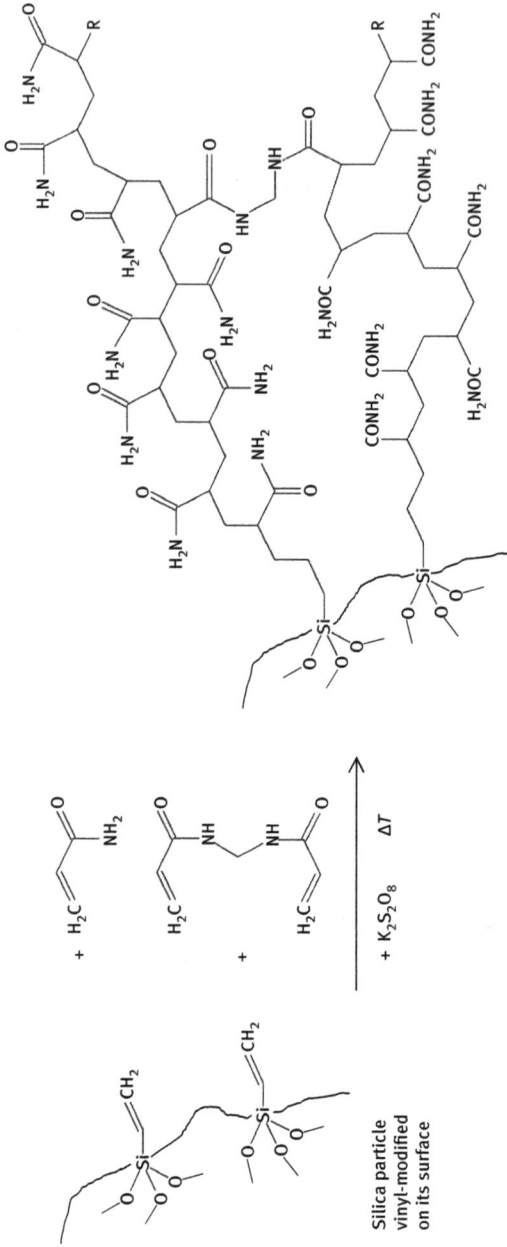

Figure 1.5: Formation of core–shell particles made of vinyl-functionalized SiO_2 and acrylamide-based SAP according to a procedure described by Maijan et al. [67].

Similarly PVA has been cross-linked with pentanedial (i.e., glutaraldehyde) [71]. Solution polymerization of sulfamic acid and acrylic acid together with starch resulted in a highly salt-tolerant SAP [72].

1.3 Biobased superabsorbers

Petrochemical-based monomers and products often lack biocompatibility, biodegradability and, as a matter of fact, they are not renewable. To step beyond these fundamental drawbacks, much attention has been paid to biopolymers in the recent years. Not only pure or just slightly modified substances were investigated for their superabsorbing properties, but conventional superabsorbing moieties were grafted onto the original biopolymers to form composite structures as well. Concise reviews have recently been published to present the numerous directions evolving [3, 73–75]. Selected examples are discussed here.

1.3.1 Polysaccharides and grafted derivatives

Among the polysaccharides, cellulose, starch, alginate, and chitin/chitosan play essential roles. The sources may be primary from nature but as well be recycled materials or by-products from other processes.

1.3.1.1 Cellulose

Shen et al. published a thorough review on cellulose- and chitin-based hydrogels in 2016 [73] and Ghorbani et al. provided a thorough further one on cellulose-based SAP in 2018 [74]. They contain concise overviews of the synthesis strategies, namely, chemically the radical solution and inverse suspension polymerizations similar to the acrylate-based procedures as well as physical methods using microwave and high-energy irradiation, chemical and physical (i.e., electrostatic) cross-linking options, and beneficial solvent systems. Some examples of cellulose-based SAP are treated in the following section.

Cellulose from eucalyptus pulp was oxidized with NaClO in the presence of a catalytical amount of (2,2,6,6-tetramethylpiperidin-1-yl)oxyl and ionically cross-linked. The products obtained were cellulose nanofibers with superabsorbent properties [76]. In another study, cellulose was cross-linked with glycine to form easily accessible SAP in one reaction step [77].

A vast majority of literature in this field describes composite structures made of the eponymous biopolymer grafted with other polymerizable substances derived from conventional SAP chemistry, mainly acrylic acid and MBA, employing the free-radical reaction mechanism. The cellulose can stem from suitable waste materials or side-products like agricultural waste [78–80] or pulp [81], or from distinct materials such as kenaf fiber [82], corn straw [83, 84], or it could be harvested from bacterial cultures [85, 86].

In a two-stage sequence, cellulose was first disintegrated to distinct degrees of polymerization and then grafted by acrylic acid cross-linked with MBA by free-radical copolymerization [87].

Carboxymethylcellulose was described being successfully cross-linked with citric acid to generate an SAP [88]. Conversion to an SAP was as well accomplished by reaction with acrylic acid and agar as the cross-linking agent under low-temperature plasma conditions. In this case, cellulose stemmed from bagasse pulp and was substituted to carboxymethylcellulose [81]. In another approach, cellulose originating from bacterial production was first carboxymethylated by sodium 2-chloroacetate and then grafted with oxiran-2-ylmethyl 2-methylprop-2-enoate (i.e., glycidyl methacrylate). Cross-linking of this intermediate product to form an SAP was accomplished by polyethylene glycol diacrylate with a molar mass of 400 g/mol, initiated by ammonium peroxodisulfate and catalyzed by N,N,N',N'-tetramethylethane-1,2-diamine [86].

Besides introducing carboxylic moieties in the graft part, hydroxymethylcellulose was grafted with the additional comonomer AMPS [89]. Cellulose fibers extracted from corn straw were grafted with AMPS and MBA in a sodium peroxodisulfate-initiated radical reaction [83].

A semi-interpenetrating polymer network working as an SAP for use as a fertilizer was synthesized from grafted corn straw cellulose and PVA by solution polymerization. The cellulose was extracted with nitric acid. Grafting of the cellulose was accomplished radically with acrylic acid. Ammonium polyphosphate was incorporated as a nutrient [84].

Nonalkylene-based grafting of cellulose was described using supercritical carbon dioxide. At a pressure of 200 bar and a temperature of 45 °C, carboxymethylcellulose-co-hydroxyethylcellulose reacted with carbon dioxide to form a superabsorbing material [90].

1.3.1.2 Starch

Similar to cellulose, starch can be used as the biopolymeric basis for grafting cross-linked polycarboxylic acids onto it. Starch can stem from rice [91], corn [48, 92], or maize bran [93]. In most cases, the grafts consisted of acrylic acid and MBA.

Maize bran, which consists of majorly starch and some cellulose, was copolymerized with acrylic acid and MBA. The reaction was initiated by UV irradiation and a composition of 2,2-dimethoxy-2-phenylacetophenone and ammonium peroxodisulfate as a radical initiator system [93]. Alternatives to acrylic acid/MBA introducing carboxylate moieties include itaconic acid and citric acid as monomer and cross-linker, respectively [94]. AMPS was used with starch as a grafting comonomer together with acrylic acid, whereby MBA served as the cross-linker, to introduce sulfonate in addition to the carboxylate groups. Salt tolerance and reswelling capability were enhanced when compared with purely carboxylate-type grafts [95]. Besides the acrylic acid in such grafting system, acrylamide was used to copolymerize with starch, AMPS, and MBA in a free-radical reaction initiated with ammonium peroxodisulfate [96].

Sulfonation of corn starch (food grade, containing 84% carbohydrate or 23% amylose, respectively) was accomplished with di- or tetra-sulfonated triazinyl compounds. The product was radically reacted with acrylonitrile, initiated by the redox system of peroxodisulfate and bisulfite, in organic solvent and then hydrolyzed with aqueous sodium hydroxide solution [97].

Similar to other types of SAP were magnetic Fe_3O_4 particles incorporated inside starch-graft (polyacrylic acid)-based SAP. The inorganic nanosized particles were present in the batch of the free-radical copolymerization and served as cores for the SAP formed, resulting in magnetic core–shell composites [98].

1.3.1.3 Alginate

Alkali alginate, mostly sodium alginate, represents a further biopolymer suitable as a substrate for SAP synthesis. Physical cross-linking of sodium alginate can electrostatically be realized by reaction with calcium chloride. The product is efficient in internal curing of concrete without impairing the mechanical strength of the hardened building material [99, 100].

Acrylic acid-graft-sodium alginate cross-linked by MBA [101], as well as accordingly cross-linked acrylamide-graft-sodium alginate [47, 55] and acrylamide/AMPS-graft-sodium alginate [47], can be prepared according to the free-radical polymerization mechanism. Initiation is possible by a thermosensitive initiator such as a peroxodisulfate [47, 101] or gamma radiation emitted from ^{60}Co [102]. Ampholytic SAP was synthesized from sodium alginate, acrylic acid, trimethyl(prop-2-enyl)azanium chloride (i.e., allyltrimethylammonium chloride), and MBA by free-radical copolymerization to form sodium alginate-graft-(polyacrylic acid-*co*-allyltrimethylammonium chloride) [103].

Prior to grafting, alginate can be methacrylated by the reaction of its hydroxyl moieties with methacrylic anhydride. This modified alginate features C=C double bonds, which are the anchor groups in radical copolymerizations with acrylic acid, acrylamide, and mixtures thereof [104, 105], as well as AMPS, acrylamide, and mixtures thereof [106], or *N*-[3-(dimethylamino)propyl]-2-methylprop-2-enamide [105]. Dimethylaminoethyl methacrylate or dimethylaminopropyl methacrylamide may be used correspondingly, to introduce amino functionalities [107]. Such

SAP are very promising materials for application in concrete, not at least due to their pronounced pH-responsive swelling ability for self-sealing and triggering self-healing of cracks [105, 107].

1.3.1.4 Chitin and chitosan

Shen et al. [73] concisely covered the chemistry of chitin- and chitosan-based SAP. The linear polysaccharides must be cross-linked physically or chemically to generate three-dimensional hydrogel structures. Exemplary molecules for cross-linking chitin, as well as cellulose, include butane-1,2,3,4-tetracarboxylic acid, oxolane-2,5-dione (i.e., succinic anhydride), 2-hydroxypropane-1,2,3-tricarboxylic acid (i.e., citric acid), 2-(chloromethyl)oxirane (i.e., epichlorohydrin), 2-[2-[2-(oxiran-2-ylmethoxy)ethoxy]ethoxymethyl]oxirane (i.e., diethylene glycol diglycidyl ether), and 1-ethenylsulfonylethene (i.e., divinyl sulfone).

The deacetylated form of chitin, denominated as chitosan, is a further worthwhile polysaccharide educt for synthesis of SAP. Chemical cross-linkers for chitosan include propanedial, pentanedial, hexamethylene-1,6-di-(aminocarboxysulfonate), and methyl (1R,4aS,7aS)-1-hydroxy-7-(hydroxymethyl)-1,4a,5,7a-tetrahydrocyclopenta[c]pyran-4-carboxylate (i.e., genipin). Alternatively, irradiation can be applied for cross-linking, which saves the use of chemical substances, for example, food applications of the SAP products [73].

Some distinct examples for chitosan-based hydrogels may be explained in detail. Chitosan can be methacrylated and this primary product be reacted with dimethylaminoethyl methacrylate or dimethylaminopropyl methacrylamide, to form superabsorbers [107]. Carboxymethyl-modified chitosan was cross-linked by glutaraldehyde. Due to the weak mechanical stability, it was reinforced with cellulose nanofibers [108]. In a comparable manner, carboxymethylated chitosan was reacted with cellulose aldehyde, the latter originating from bleached kraft pulp to form a fully biobased SAP without any by-products except for water [109]. Chitosan with a degree of 85% of deacetylation was reacted with gelatin to generate a composite foam that behaves as an SAP. A ternary solvent system of dioxane, acetic acid, and water was used at elevated temperature. Finally, the produce was frozen at a temperature of −196 °C and lyophilized for foam formation and removal of the solvents, respectively [110].

A mineral-modified composite SAP was synthesized from chitosan, basalt particles, acrylic acid, acryl amide, AMPS, and the cross-linker MBA using microwave irradiation to generate chitosan-graft-poly(AMPS-co-acrylic acid-co-acrylamide)/ground basalt (Figure 1.6). Interestingly, the basalt significantly improved the antibacterial activity when compared with the mere polymer [111].

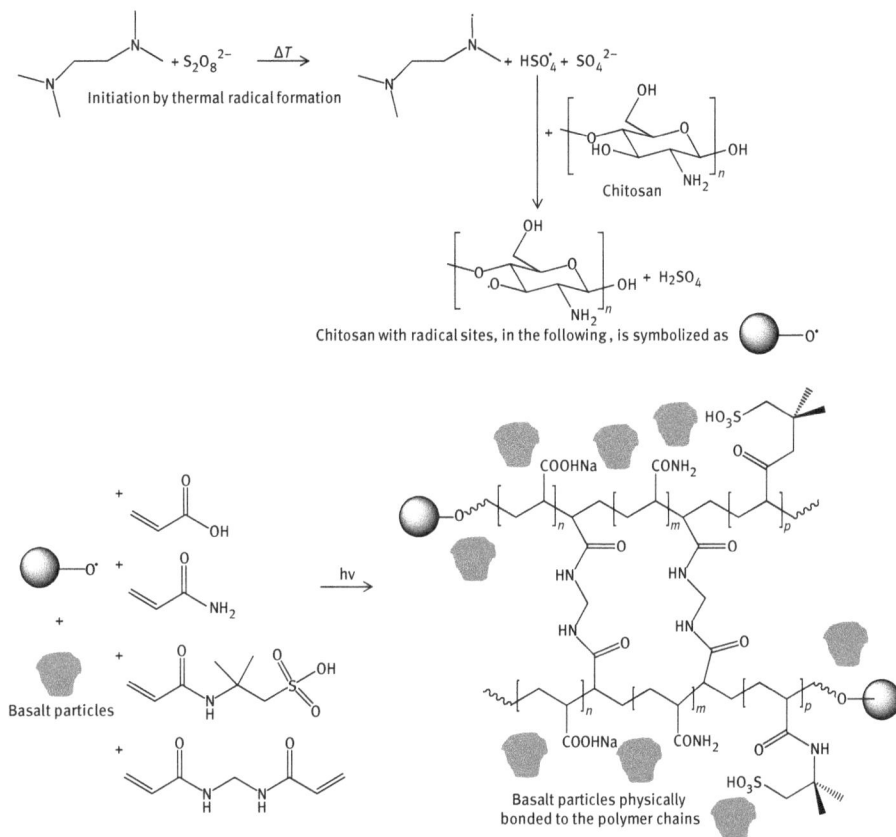

Figure 1.6: Reaction scheme to synthesize chitosan-graft-poly(2-acrylamido-2-methylpropanesulfonic acid-*co*-acrylic acid-*co*-acrylamide)/ground basalt, radical formation at the chitosan first, followed by irradiation-initiated grafting and incorporation of basalt as suggested by Atassi et al. [111].

1.3.1.5 Further saccharide-based SAP

Paste-like mixtures of predissolved carboxymethylcellulose and starch from potatoes were irradiated by gamma rays using ^{60}Co. The cross-linking resulted in gel-stiff solutions, which were dried to obtain a superabsorbent solid product [112].

K-carrageenan, which consists mainly of copoylmerized α-1,3- and β-1,4-linked sulfate esters of D-galactose and 3,6-anhydro-D-galactose, respectively, was grafted with acrylic acid, cross-linked by MBA according to the free-radical polymerization mechanism. The product featured good ability to mitigate autogenous shrinkage of cement-based building composites [113].

Equivalent to alginate and chitosan, agarose can be methacrylated. The product of this reaction can further be treated with dimethylaminoethyl methacrylate or dimethylaminopropyl methacrylamide to form according to superabsorbers [107].

Polysaccharide-based superabsorbers may suffer mechanical instability. To overcome this issue, powdery ingredients were introduced in the synthesis process, for example, kaolin [80], char [92], or rice husk ash [114].

Besides polysaccharides, monosaccharides can also be used as the educt for SAP synthesis. Glucose was metabolized via mevalonate to isoprene carboxylic acid, which was then radically solution-polymerized to a cross-linked network. An extra prepared additional cross-linker was incorporated as well, by name butane-1,4-bisisoprenecarboxylic acid ester [115].

1.3.2 Protein-based SAP

Wheat flour is the basis of biodegradable SAP based on proteins. After dissolution, it was radically grafted with acrylamide using gamma rays [116]. In another procedure, the protein chains (obtained from wheat gluten) were cross-linked by genipin (methyl (1R,2R,6S)-2-hydroxy-9-(hydroxymethyl)-3-oxabicyclo[4.3.0]nona-4,8-diene-5-carboxylate) via lysine residues. For mechanical stabilization of the resulting foam-like structure, cellulose nanofibers were added, which are supposed to covalently bond to residues of the genipin via carboxylate groups (Figure 1.7) [117]. An amphoteric SAP with carboxylic and ammonium functionalities was prepared from leather waste, acrylic acid, dimethyl-bis(prop-2-enyl)azanium chloride (i.e., dimethyldiallylammonium chloride) and MBA. Proteins contained in the organic waste material were considered to be the essential substance for SAP formation [118]. Crucial synthesis steps and a characteristic pattern of the product are sketched in Figure 1.8.

1.3.3 Other natural or nature-derived educts

From pectin, SAP were synthesized. Radical copolymerization for grafting was accomplished with acrylic acid and MBA [119, 120]. Exemplarily, the pectin can be extracted from passion fruit peel [119].

Lignin was used as a further biobased educt for SAP synthesis. Alkali lignin, which had been reacted with ethylene glycol, as well as lignin nanoparticles were grafted by free-radical copolymerization with acrylic acid and MBA [121].

In another example, sodium humate was added to a conventional free-radical synthesis scheme of acrylic acid-co-AMPS SAP, cross-linked with MBA. The swelling ability in saline solutions increased when compared with humate-free reference products [122].

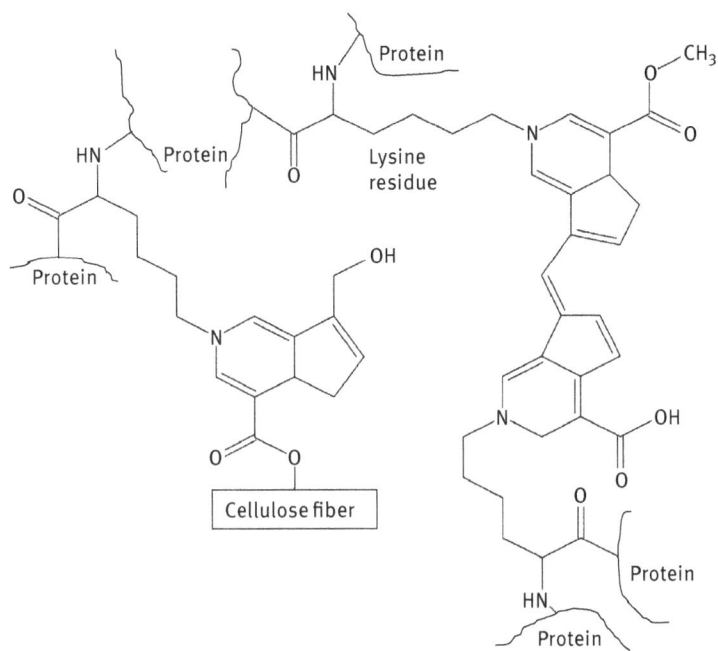

Figure 1.7: Constitutional formula of protein-based superabsorber prepared with genipin as a cross-linker and reinforced with cellulose nanofibers as it was proposed by Capezza et al. [117]; interestingly, the publication cited is inconsistent within itself with respect to the bond properties in the cross-linking units stemming from the genipin – the representation at hand covers any of the suggested ones.

Tannic acid was researched as a biobased cross-linker to replace petrochemical substances. It was modified with alkene and epoxy groups to act as either an internal or an external cross-linker, respectively (Figure 1.9) [123].

Apart from well-defined biobased substances, waste materials have been investigated as a resource for SAP, which have hardly been treated before the synthesis. The major chemicals contained in such mixtures are supposed to be polysaccharides and proteins that are cross-linked in the respective reactions. In this sense, porcine plasma protein was reacted with glycerol propane-1,2,3-triol [124]. Wu et al. reacted soybean residue, which is mainly cellulose, with acrylic acid, poly(ethylene glycol), urea, and kaolin to an SAP produce by irradiation with UV light [80]. Agricultural waste soybean dregs were grafted with acrylic acid, cross-linked with MBA, by using UV radiation, to form a superabsorbent produce. The soybean dregs contained 26.1% crude cellulose, 12.8% crude protein, 9.38% moisture, 2.77% crude fat, and unspecifically bound nitrogen in the elemental proportion of 2.05% [78].

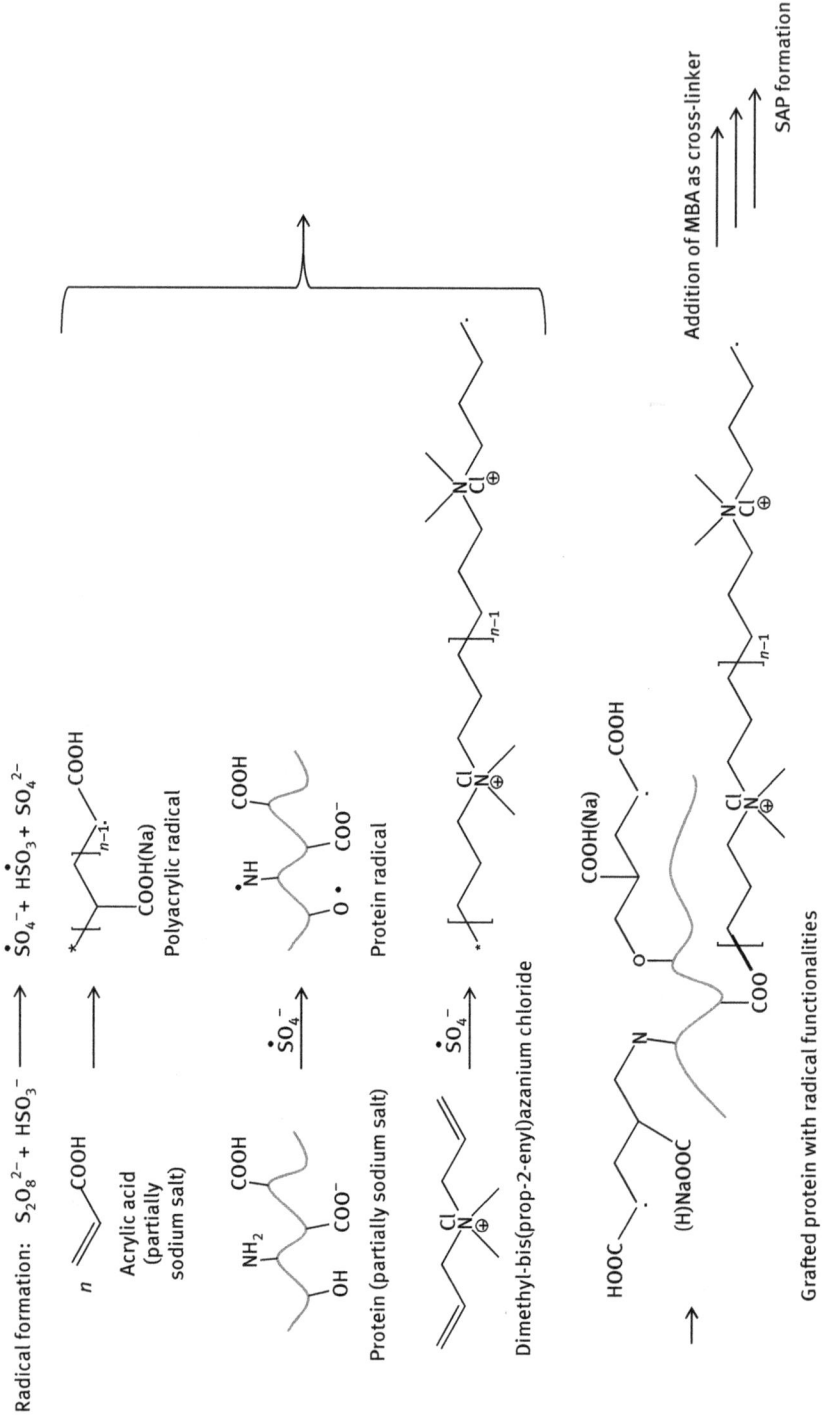

Figure 1.8: Reaction scheme for synthesis of amphoteric SAP based on leather waste proteins as outlined by Xu et al. [118].

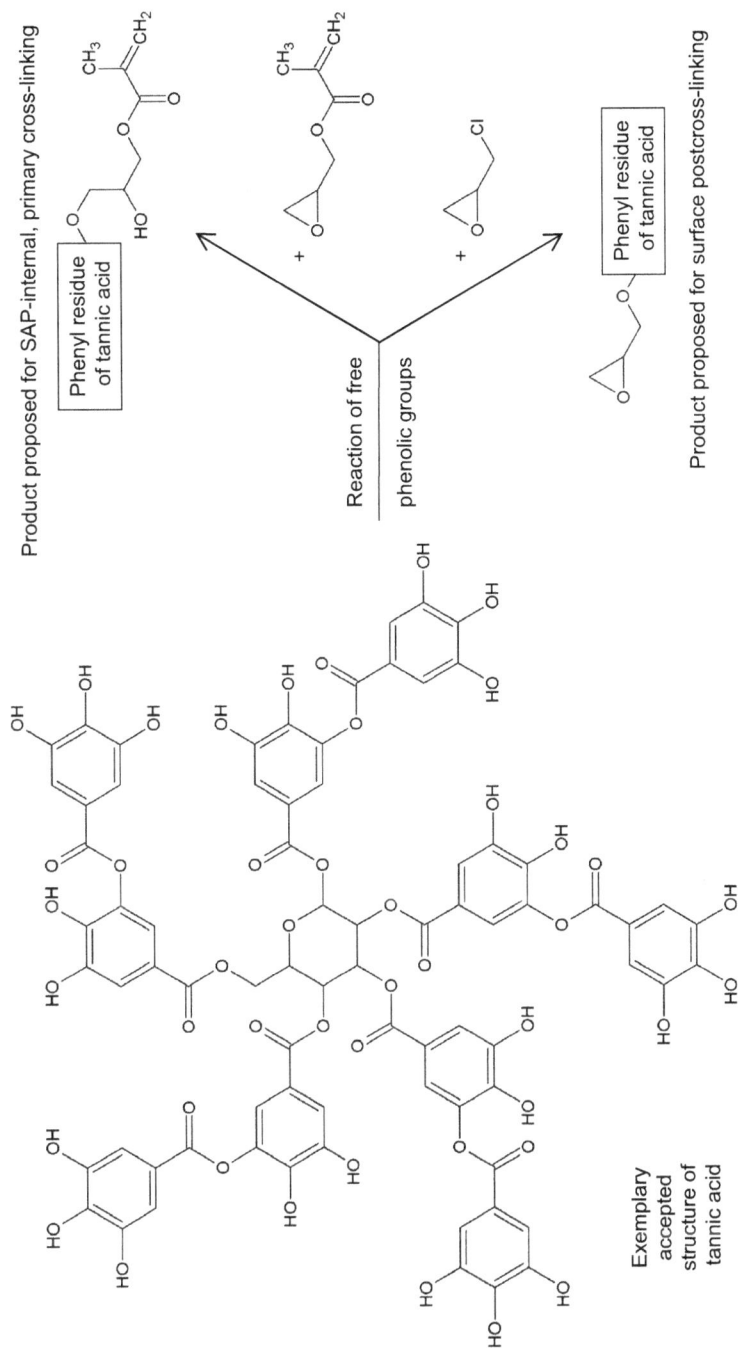

Figure 1.9: Alkene- or epoxy-functionalization of tannic acid for use as a biobased cross-linking agent in SAP synthesis, both as a primary cross-linker and for the purpose of surface postcross-linking suggested and structurally characterized by Dabbaghi et al. [123].

Costream proteins from potato processing, which cannot be used in the food production, were acylated with succinic anhydride, citric acid, 1,2,3,4-butanetetracarboxylic acid, EDTA, and EDTA-dianhydride, respectively, to obtain a superabsorbing gel product [125].

1.4 Summary and conclusion

SAP developments have mainly been driven by requirements stemming from applications in sanitary products. Predominantly, SAPs are based on acrylate chemistry and free-radical copolymerization. For distinct purposes, comonomers can be integrated to obtain specific effects, that is, sulfonic groups for SAP in cement-based building materials or biologically active moieties. After synthesis, the primary products are processed by elaborate postpolymerization treatments, being surface post-cross-linking of utmost importance to minimize the so-called gel-blocking effect in SAP beds or the stickiness of powdery SAP products.

A dynamically emerging field of SAP research is the use of biopolymers or bio-derived main monomers and cross-linkers. The natural substances are either chemically modified just to a minor extent or they are heavily processed by grafting conventional acrylate-based superabsorbing network structures onto them. This opens up a broad field for chemical synthesis that ranges from pure biobased SAP to hybrid products from petrochemical and biogenic resources. Hereby, polysaccharides represent a most intensely researched class of substances.

References

[1] Richter A., Paschew G., Klatt S., Lienig J., Arndt K.F., Adler H.J.P. Review on hydrogel-based pH sensors and microsensors. Sensors 2008, 8(1), 561–581.
[2] Bauer S., Baumann K., Toennessen M., Bauduin C., Biel M., Daniel T. Superabsorber mixtures. WO 2019/137833 A1, 18.07.2019.
[3] Santos R.V.A., Costa G.M.N., Pontes K. Development of tailor-made superabsorbent polymers: review of key aspects from raw material to kinetic model. J Polym Environ 2019, 27, 1861–1877.
[4] Tsukabimoto T., Shimomura T., Irie Y., Masuda Y. Absorbent resin composition and process for producing same. US 4,286,082, 25.08.1981.
[5] Dahmen K., Mertens R. Pulverförmige, vernetzte, wässrige Flüssigkeiten sowie Blut absorbierende Polymere, Verfahren zu ihrer Herstellung und ihre Verwendung als Absorptionsmittel in Hygieneartikeln. DE 4020780 C1, 29.06.1990.
[6] Chmelir M., Pauen J. Process for the continuous production of polymers and copolymers of water-soluble monomers. US 4,857,610, 15.08.1989.
[7] Braig V., de Marco M., Stösser M., Pakusch J. Method for the production of superabsorbers. WO 2008/046841 A1, 24.04.2008.

[8]   Stephan O., Barthel H., Grünewald G. Vorrichtung zum Kühlen und Rückbefeuchten von Superabsorberpartikeln. DE 10 2016 208652 A1, 23.11.2017.
[9]   Stephan O., Grünewald G., Weismantel M., Possemiers K., Peterson M., Funke R., Bertha S., Catternan M. Method for the production of superabsorbers. WO 2019/076682 A1, 25.04.2019.
[10] Anil I., Gunday S.T., Alagha O., Bozkur A. Synthesis, characterization, and swelling behaviors of poly(acrylic acid-co-acrylamide)/pozzolan superabsorbent polymers. J Polym Environ 2019, 27, 1086–1095.
[11] Farzanian K., Ghahremaninezhad A. On the effect of chemical composition on the desorption of superabsorbent hydrogels in contact with a porous cementitious material. Gels 2018, 4, 70.
[12] Safaei F., Khalilii S., Khorasani S.N., Neisiany R.E. Preparation of an acrylic acid-based superabsorbent composite: investigations of synthesis parameters. Chem Pap 2020, 74, 939–949.
[13] Obayashi S., Nakamura M., Fujiki K., Yamamoto T. Alkali metal acrylate or ammonium acrylate polymer excellent in salt solution-absorbency and process for producing the same. US 4,340,706, 20.07.1982.
[14] Daamen K., Retorofu H. Manufacture of high water absorption polymer. JPS6438406 (A), 08.02.1989.
[15] Von der Heydt M.A., Krueger M., Stanitzek U. Method for the discontinuous production of superabsorber particles by polymerizing an aqueous monomer solution dispersed in a hydrophobic solvent. WO 2018/202489 A1, 08.11.2018.
[16] Von der Heydt M.A., Stanitzek U., Krueger M. Method for the discontinuous production of superabsorber particles by polymerizing an aqueous monomer solution dispersed in a hydrophobic solvent. WO 2018/202490 A1, 08.11.2018.
[17] Tang Y., Tang H., Wang F., Guan C., Zhu L. Synthesis and swelling behavior of poly(acrylic acid)/graphite oxide superabsorbent composite. Polym Sci, Series B 2019, 61(4), 471–478.
[18] Li J., Zhang K., Zhang M., Fang Y., Chu X., Xu L. Fabrication of a fast-swelling superabsorbent resin by inverse suspension polymerization. J Appl Polym Sci 2018, 135(15), 46142.
[19] Glienke P.O., Glienke I.M. Hydrogelkomposit-Suspension. DE 20 2011 002 784 U1, 21.04.2011.
[20] Glienke P.O., Glienke I.M. Hydrogelkomposit für Suspensionsapplikationen. DE 20 2011 003 679 U1, 22.06.2011.
[21] Parrish M., Dujardin R. Methods of increasing crop yield and controlling the growth of weeds using a polymer composite film. US 2011/0152100 A1, 23.06.2011.
[22] Loewe-Fürstenberg A. Düngemittel-Agglomerat mit Superabsorber. DE 20 2012 001 243 U1, 26.04.2012.
[23] Gross H.J., Wack H., Althaus W. Diaper with a heat-sensitive or pH-sensitive superabsorber. WO 99/03435, 28.01.1999.
[24] Bennett A.K., Herfert N., Todd M.A. Superabsorber that comprises at least one sterically hindered monovalent and/or polyvalent phenol. WO 2005/054356 A1, 16.06.2005.
[25] Torii K., Kitayama T., Harada N. Water-absorbing agent and production process therefore, and water-absorbent structure. US 2006/0229413 A1, 12.10.2006.
[26] Nakamura M., Miyake K., Iwamura T., Watanabe Y. Water-absorbent resin composition, method of manufacturing the same, and absorbent article. WO 2007/072969 A1, 28.06.2007.
[27] Fujimaru H., Goto E., Ishizaki K., Motoyama A. Method for producing polyacrylic acid (salt) water-absorbent resin. US 2007/0232760 A1, 04.10.2007.
[28] Torii K., Adachi Y., Kobayashi M., Watanabe Y., Ikeuchi H., Kitayama T. Water-absorbing agent and production method thereof. US 2008/0221229 A1, 11.09.2008.

[29] Champ S., Herrlich-Loos M. Superabsorber comprising a virus-inhibiting additive. WO 2009/040358 A2, 02.04.2009

[30] Flohr A., Lindner T., Mitsukami Y. Method of surface cross-linking superabsorbent polymer particles using ultraviolet radiation and Bronsted acids. US 2009/0131633 A1, 21.05.2009.

[31] Flohr A., Lindner T., Oliveros E., Mitsukami Y. Method of surface cross-linking superabsorbent polymer particles using ultraviolet radiation. US 2009/0137694 A1, 28.05.2009.

[32] Torii K., Kobayashi T. Water absorbent and process for production thereof. WO 2009/093708 A1, 30.07.2009.

[33] Flohr A., Lindner T., Oliveros E., Mitsukami Y. Method of surface cross-linking superabsorbent polymer particles using vacuum ultraviolet radiation. US 2009/0197985 A1, 06.08.2009.

[34] Ikeuchi H., Nagasawa M. Particulate water-absorbent polymer and production method thereof. US 2009/0270538 A1, 29.10.2009.

[35] Nagasawa M., Ikeuchi H. Particulate water absorbing agent and manufacture method of same. US 2009/0275470 A1, 05.11.2009.

[36] Torii K., Shibata M., Kimura K., Nakashima Y., Imura M., Ueda H., Wada K. Water absorbing agent, water absorbent core using the agent, and manufacturing method for water absorbing agent. US 2009/0298685 A1, 03.12.2009.

[37] Bruhns S., Daniel T., Hermeling D., Riegel U. Mixture of surface postcrosslinked superabsorbers with different surface postcrosslinking. WO 2010/057823 A1, 27.05.2010.

[38] Wada K., Nakashima Y. Water-absorbing resin composition and process for production of the same. US 2010/0240808 A1, 23.09.2010.

[39] Torii K., Kobayashi T., Adachi Y., Watanabe Y., Kitayama T. Water absorbing agent and production method thereof. US 2011/00114881 A1, 19.05.2011.

[40] Adachi Y., Torii K., Watanabe Y., Kobayashi T., Kitayama T., Suzuki S., Yoneda A. Water absorbing agent and production method thereof. US 2011/0180755 A1, 28.07.2011.

[41] Tada K., Kadonaga K., Sasabe M. Binding method of water absorbent resin. US 2011/0237739 A1, 29.09.2011.

[42] Fujino S., Nagasawa E., Matsumoto S., Ishizaki K. Polyacrylic acid (salt) type water-absorbent resin and process for production of same. EP 2 395 029 A1, 12.12.2011.

[43] Langlotz J.K., Friedrich S., Hesse C. Pulverförmiger Beschleuniger. WO 2012/072466 A1, 07.06.2012.

[44] Funkhouser G.P., Benkley J.R. Delayed, swellable particles for prevention of fluid migration through damaged cement sheaths. WO 2013/062700 A1, 02.05.2013.

[45] Bauduin C., Daniel T. Permeable superabsorber and the production thereof. WO 2019/197194 A1, 17.10.2019.

[46] Rychter P., Rogacz D., Lewicka K., Kollár J., Kawalec M., Mosnáček J. Ecotoxicological properties of Tulipalin A-based superabsorbents versus conventional superabsorbent hydrogels. Adv Polym Technol 2019, 2947152.

[47] Rehman T.U., Bibi S., Khan M., Ali I., Shah L.A., Khan A., Ateeq M. Fabrication of stable superabsorbent hydrogels for successful removal of crystal violet from waste water. RSC Adv 2019, 9, 40051.

[48] Wang C., Bu Y., Guo S., Lu Y., Sun B., Shen Z. Self-healing cement composite: amine- and ammonium-based pH-sensitive superabsorbent polymers. Cem Concr Compos 2019, 96, 154–162.

[49] Miyajima T., Matsubara Y., Komatsu H., Miyamoto M., Suzuki K. Development of a superabsorbent polymer using iodine transfer polymerisation. Polym J 2019. https://doi.org/10.1038/s41428-019-0292-2.

[50] Dairoku Y., Fujimaru H., Ishizaki K. Particulate water absorbing agent including polyacrylic acid (polyacrylate) based water absorbing resin as a principal component, method for production thereof, water-absorbent core and absorbing article in which the particulate water absorbing agent is used. US 2009/0318885 A1, 24.12.2009.

[51] Herbert N., Parchana B., Kaenthong S., Bauer S., Daniel T., Baumann K. Process for producing superabsorbents. WO 2019/201669 A1, 24.10.2019.

[52] Ghasri M., Bouhendi H., Kabiri K., Zohuriaan-Mehr M.J., Karami Z., Omidian H. Superabsorbent polymers achieved by surface cross linking of poly(sodium acrylate) using microwave method. Iran Polym J 2019, 28, 539–548.

[53] Mojarad-Jabali S., Kabiri K., Karami Z., Mastropietro D.J., Omidian H. Surface cross-linked SAPs with improved swollen gel strength using diol compounds. J Macromol Sci Part A: Pure Appl Chem 2020, 57(1), 62–71.

[54] Kim Y.J., Hong S.J., Shin W.S., Kwon Y.R., Lim S.H., Kim H.C., Kim J.S., Kim J.W., Kim D.H. Preparation of a biodegradable superabsorbent polymer and measurements of changes in absorption properties depending on the type of surface-crosslinker. Polym Adv Technol 2020, 31, 273–283.

[55] Jiang Z., Cao X., Guo L. Synthesis and swelling behavior of poly (acrylic acid-acryl amide-2-acrylamido-2-methyl-propansulfonic acid) superabsorbent copolymer. J Pet Explor Prod Technol 2017, 7, 69–75.

[56] Wu X., Huang X., Zhu Y., Li J., Hoffmann M.R. Synthesis and application of superabsorbent polymer microspheres for rapid concentration and quantification of microbial pathogens in ambient water. Sep Purif Technol 2020, 239, 116540.

[57] Ashkani M., Bouhendi H., Kabiri K., Rostami M.R. Synthesis of poly (2-acrylamido-2-methyl propane sulfonic acid) with high water absorbency and absorption under load (AUL) as concrete grade superabsorbent and its performance. Constr Build Mater 2019, 206, 540–551.

[58] Hartnagel K., Grieger T., Daniel T., Schroeder M., Morano M., Greene N.T. Method for the production of superabsorbers. WO 2018/029045 A1, 25.02.2018.

[59] Khakpour H., Abdollahi M. Synthesis, characterization and rheological properties of acrylamide/ acidic monomer/ N-(4-ethylphenyl) acrylamide terpolymers as pH-responsive hydrogels and nanogels. Polym-Plast Technol Mat 2020, 59(4), 441–455.

[60] Strassburg A., Petranowitsch J., Paetzold F., Krumm C., Peter E., Meuris M., Köller M., Till J.C. Cross-linking of a hydrophilic, antimicrobial polycation toward a fast-swelling, antimicrobial superabsorber and interpenetrating hydrogel networks with long lasting antimicrobial properties. Appl Mater Interfaces 2019, 9, 36573–36582.

[61] Cavallo A., Madaghiele M., Masullo U., Lionetto M.G., Sannino A. Photo-crosslinked poly (ethylene glycol) diacrylate (PEGDA) hydrogels from low molecular weight prepolymer: swelling and permeation studies. J Appl Polym Sci 2017, 134(2), 44380.

[62] Voll N.E., Stephan O., Schumann P., Grünewald G. Method for drying superabsorbers. WO 2016/128337 A1, 28.08.2016.

[63] Vivekh P., Islam M.R., Chua K.J. Experimental performance evaluation of a composite superabsorbent polymer coated heat exchanger based air dehumidification system. Appl Energy 2020, 260, 114256.

[64] Jeddi M.K., Laitinen O., Liimatainen H. Magnetic superabsorbents based on nanocellulose aerobeads for selective removal of oils and organic solvents. Mater Des 2019, 183, 108115.

[65] Moini N., Zohuriaan-Mehr M.J., Kabiri K., Khonakdar H.A. "Click" on SAP: superabsorbent polymer surface modification via CuAAC reaction towards antibacterial activity and improved swollen gel strength. Appl Surf Sci 2019, 487, 1131–1144.

[66] Ghomi E.R., Khorasani S.N., Kichi M.K., Dinari M., Ataei S., Enayati M.H., Koochaki M.S., Neisiany R.E. Synthesis and characterization of TiO_2/acrylic acid-co-2-acrylamido-2-methyl

propane sulfonic acid nanogel composite and investigation its self-healing performance in the epoxy coatings. Colloid Polym Sci 2020, 298, 213–223.

[67]   Maijan P., Chantarak S. Synthesis and characterization of highly durable and reusable superabsorbent core-shell particles. Polym Eng Sci 2020, 60, 306–313.

[68]   Kanellopoulou I., Karaxi E.K., Karatza A., Kartsonakis I.A., Charitidis C. Hybrid superabsorbent networks (SAPs) encapsulated with SiO_2 for structural applications. MATEC Web Conf 2018, 188, 01025.

[69]   Zhang M., Zhang S., Chen Z., Wang M., Cao J., Wang R. Preparation and characterization of superabsorbent polymers based on sawdust. Polymers 2019, 11, 1891.

[70]   Li Z., Wang D., Bai H., Zhang S., Ma P., Dong W. Photo-crosslinking strategy constructs adhesive, superabsorbent, and tough PVA-based hydrogel through controlling the balance of cohesion and adhesion. Macromol Mater Eng 2020, 305, 1900623.

[71]   Thakur K., Rajhans A., Kandasubramanian B. Starch/PVA hydrogels for oil/water separation. Environ Sci Pollut Res 2019, 26, 32013–32028.

[72]   Zhao C., Zhang M., Liu Z., Guo Y., Zhang Q. Salt-tolerant superabsorbent polymer with high capacity of water-nutrient retention derived from sulfamic acid-modified starch. ACS Omega 2019, 4, 5923–5930.

[73]   Shen X., Shamshina J.L., Berton P., Gurau G., Rogers R.D. Hydrogels based on cellulose and chitin: fabrication, properties and applications. Green Chem 2016, 18, 53–75.

[74]   Ghorbani S., Eyni H., Bazaz S.R., Nazari H., Asl L.S., Zaferani H., Kiani V., Mehrizi A.A., Soleimani M. Hydrogels based on cellulose and its derivatives: applications, synthesis, and characteristics. Polym Sci Ser A 2018, 60(6), 707–722.

[75]   Mignon A., De Belie N., Dubruel P., Van Vlierberghe S. Superabsorbent polymers: a review on the characteristics and applications of synthetic, polysaccharide-based, semi-synthetic and 'smart' derivatives. Eur Polym J 2019, 117, 165–175.

[76]   Patiño-Masó J., Serra-Parareda F., Tarrés Y., Mutjé P., Espinach F.X., Delgado-Aguilar M. TEMPO-oxidized cellulose nanofibers: a potential bio-based superabsorber of diaper production. Nanomaterials 2019, 9, 1272.

[77]   Palantöken S., Bethke K., Zivanovic V., Kalinka G., Kneipp J., Rademann K. Cellulose hydrogels physically crosslinked by glycine: synthesis, characterization, thermal and mechanical properties. J Appl Polym Sci 2020, 48380.

[78]   Fan Y., Zhang M., Shangguan L. Synthesis of a novel and salt sensitive superabsorbent hydrogel using soybean dregs by UV-irradiation. Materials 2018, 11, 2198.

[79]   Li S., Chen G. Agricultural waste-derived superabsorbent hydrogels: preparation, performance and socioeconomic impacts. J Clean Prod 2020, 251, 119669.

[80]   Wu C., Wang L., Kang H., Dan Y., Tian D., Zheng Y. UV-light irradiation preparation of soybean residue-based hydrogel composite from inorganic/organic hybrids for degradable slow-release N-fertilizer. Res Chem Intermed 2020, 46, 1437–1451.

[81]   Huang L., Zhao H., Xu H., Qi M., Yi T., Huang C., Wang S., An S., Li C. Kinetic model of a carboxymethylcellulose-agar hydrogel for long-acting and slow-release of chlorine dioxide with a modification of Fick's diffusion law. Biorecourses 2019, 14(4), 8821–8834.

[82]   Zakaria M.E.B.T., Jamari S.S.B., Ghazali S. Synthesis of superabsorbent carbonaceous kenaf fibre filled polymer using inverse suspension polymerisation. J Mech Eng Sci 2017, 11(3), 2794–2800.

[83]   Cheng W.M., Hu X.M., Wang D.M., Liu G.H. Preparation and characteristics of corn straw-co-AMPS-co-AA superabsorbent hydrogel. Polymers 2015, 7(11), 2431–2445.

[84]   Wang W., Yang S., Zhang A., Yang Z. Preparation and properties of novel corn straw cellulose-based superabsorbent with water-retaining and slow-release functions. J Appl Polym Sci 2020, 48951.

[85] Luo M.T., Li H.L., Huang C., Zhang H.R., Xiong L., Chen X.F., Chen X.D. Cellulose-based absrobent production from bacterial cellulose and acrylic acid: synthesis and performance. Polymers 2018, 10(7), 702.
[86] Chaiyasat A., Jearanai S., Christopher L.P., Alam M.N. Novel superabsorbent polymers from bacterial cellulose. Polym Int 2019, 68(1), 102–109.
[87] Li H.X., Wang Q., Zhang L., Tian X., Cao Q., Jin L. Influence of the degree of polymerisation on the water absorption performance of hydrogel and adsorption kinetics. Polym Degrad Stab 2019, 168, 108958.
[88] Lacoste C., Lopez-Cuesta J.M., Bergeret A. Development of a biobased superabsorbent polymer from recycled cellulose for diapers applications. Eur Polym J 2019, 116, 38–44.
[89] Cheng S., Liu X., Zhen Y., Lei Z. Preparation of superabsorbent resin with fast water absorption rate based on hydroxymethyl cellulose sodium and its application. Carbohydr Polym 2019, 225, 115214.
[90] Baldino L., Zuppolini S., Cardea S., Diodato L., Borriello A., Reverchon E., Nicolais L. Production of biodegradable superabsorbent aerogels using a supercritical CO_2 assisted drying. J Supercrit Fluids 2020, 156, 104681.
[91] Zhang Q., Wang Z., Zhang C., Aluko R.E., Yuan Y., Ju X., He R. Structural and functional characterization of rice starch-based superabsorbent polymer materials. Int J Biolog Macromolecules 2020, 153, 1291–1298.
[92] Motamedi E., Motesharezedeh B., Shirinfekr A., Samar S.M. Synthesis and swelling behavior of environmentally friendly starch-based superabsorbent hydrogels reinforced with natural char nano/micro particles. J Environ Chem Eng 2020, 8, 103583.
[93] Zhang M., Cheng Z., Zhao T., Liu M., Hu M., Li J. Synthesis, characterization, and swelling behaviors of salt-sensitive maize bran-poly(acrylic acid) superabsorbent hydrogel. J Agric Food Chem 2014, 62, 8867–8874.
[94] Duquette D., Nzediegwu C., Portillo-Perez G., Dumont M.J., Prasher S. Eco-friendly synthesis of hydrogels from starch, citric acid, and itaconic acid: Swelling capacity and metal chelation properties. Starch 2019, 1900008.
[95] Meng Y., Ye L. Synthesis and swelling property of superabsorbent starch grafted with acrylic acid/2-acrylamido-2-methyl-1-propanesulfonic acid. J Sci Food Agric 2017, 97(11), 3831–3840.
[96] Liu R., Sun Z., Ding Q., Chen P., Chen K. Mitigation of early-age cracking of concrete based on a new gel-type superabsorbent polymer. ASCE J Mater Civ Eng 2017, 29(10), 04017151.
[97] Zain G., Nada A.A., El-Sheick M.A., Attaby F.A., Waly A.I. Superabsorbent polymer based on sulfonated-starch for improving water and saline absorbency. Int J Biol Macromol 2018, 115, 61–68.
[98] Saberi A., Alipour E., Sadeghi M. Superabsorbent magnetic Fe_3O_4-based starch-poly (acrylic acid) nanocomposite hydrogel for efficient removal of dyes and heavy metal ions from water. J Polym Res 2019, 26, 271.
[99] Mignon A., Snoeck D., D'Halluin K., Balcaen L., Vanhaecke F., Dubruel P., Van Vlierberghe S., De Belie N. Alginate biopolymers: counteracting the impact of superabsorbent polymers on mortar strength. Constr Build Mater 2016, 110, 169–174.
[100] Hu M., Guo J., Du J., Liu Z., Li P., Ren X., Feng Y. Development of Ca^{2+}-based, ion-responsive superabsorbent hydrogel for cement applications: self-healing and compressive strength. J Colloid Interface Sci 2019, 538, 397–403.
[101] Thakur S., Arotiba O.A. Synthesis, swelling and adsorption studies of a pH-responsive sodium alginate-poly(acrylic acid) superabsorbent hydrogel. Polym Bull 2018, 75, 4587–4606.
[102] Ghobashy M.M., Bassioni G. pH stimuli-responsive poly(acrylamide-co-sodium alginate) hydrogels prepared by γ-radiation for improved compressive strength of concrete. Adv Polym Technol 2018, 37(6), 2123–2133.

[103] Tang S., Zhao Y., Wang H., Wang Y., Zhu H., Chen Y., Chen S., Jin S., Yang Z., Li P., Li S. Preparation of the sodium alginate-g-(polyacrylic acid-co-allyltrimethylammonium chloride) polyampholytic superabsorbent polymer and its dye adsorption property. Mar Drugs 2018, 16, 476.

[104] Mignon A., Vermeulen J., Graulus G.J., Martins J., Dubruel P., De Belie N., Van Vlierberghe S. Characterization of methacrylated alginate and acrylic monomers as versatile SAPs. Carbohydr Polym 2017, 168, 44–51.

[105] Mignon A., Vermeulen J., Snoeck D., Dubruel P., Van Vlierberghe S., De Belie N. Mechanical and self-healing properties of cementitious materials with pH-responsive semi-synthetic superabsorbent polymers. Mater Struct 2017, 50, 238.

[106] Mignon A., Devisscher D., Graulus G.J., Stubbe B., Martins J., Dubruel P., De Belie N., Van Vlierberghe S. Combinatory approach of methacrylated alginate and acid monomers for concrete applications. Carbohydr Polym 2017, 155, 448–455.

[107] Mignon A., Devisscher D., Vermeulen J., Vagenende M., Martins J., Dubruel P., De Belie N., Van Vlierberghe S. Characterization of methacrylated polysaccharides in combination with amine-based monomers for application in mortar. Carbohydr Polym 2017, 168, 173–181.

[108] Bahrami B., Behzad T., Zamani A., Heidarian P., Nasri-Nasrabadi B. Synthesis and characterization of carboxymethyl chitosan superabsorbent hydrogels reinforced with sugarcane bagasse cellulose nanofibers. Mater Res Express 2019, 6, 065320.

[109] Alam M.N., Christopher L.P. Natural cellulose-chitosan cross-linked superabsorbent hydrogels with superior swelling properties. ACS Sustain Chem Eng 2018, 6(7), 8736–8742.

[110] Peng J., Wang X., Lou T. Preparation of chitosan/gelatin composite foam with ternary solvents of dioxane/acetic acid/water and its water absorption capacity. Polym Bull 2019. https://doi.org/10.1007/s00289-019-03016-2.

[111] Atassi Y., Said M., Tally M., Kouba L. Synthesis and characterization of chitosan-g-poly (AMPS-co-AA-co-AM)/ground basalt composite hydrogel: antibacterial activity. Polym Bull 2019. https://doi.org/10.1007/s00289-019-03017-1.

[112] Fekete T., Borsa J., Takács E., Wojnárovits. Synthesis of carboxymethylcellulose/starch superabsorbent hydrogels by gamma-irradiation. Chem Cent J 2017, 11, 46.

[113] Aday A.A., Osio-Norgaard J., Foster K.E.O., Srubar III W.V. Carrageenan-based superabsorbent biopolymers mitigate autogenous shrinkage in ordinary portland cement. Mater Struct 2018, 51, 37.

[114] De Vasconcelos M.C., Gomes R.F., Sousa A.A.L., Moreira F.J.C., Rodrigues F.H.A., Fajardo A.R., Pinheiro Neto L.G. Superabsorbent hydrogel composite based on starch/rice husk ash as a soil conditioner in melon (cucumis melo l) seedling culture. J Polym Environ 2020, 28, 131–140.

[115] Fahnhorst G.W., Hoye T.R. Superabsorbent poly(isoprene carboxylate) hydrogels from glucose. ACS Sustain Chem Eng 2019, 7, 7491–7495.

[116] Khozemy E.E., Nasef S.M., Mohamed T.M. Radiation synthesis of superabsorbent hydrogel (wheat flour/acrylamide) for removal of mercury and lead ions from waste solutions. J Inorg Organomet Polym Mater 2019. https://doi.org/10.1007/s10904-019-01350-6.

[117] Capezza A.J., Wu Q., Newson W.R., Olsson R.T., Espuche E., Johansson E., Hedneqvist S. Superabsorbent and fully biobased protein foams with a natural cross-linker and cellulose nanofibers. ACS Omega 2019, 4, 18257–18267.

[118] Xu S., Yin Y., Wang Y., Li X., Hu Z., Wang R. Amphoteric superabsorbent polymer based on waste collagen as loading media and safer release systems for herbicide 2, 4D. J Appl Polym Sci 2020, 48480.

[119] Carmo I.A.D., de Almeida C.A., Brandão H.M., de Oliveira L.F.C., de Souza N.L.G.D. Experimental planning applied to the synthesis of superabsorbent polymer by acrylic acid graft in pectin extracted from passion fruit peel. Mater Res Express 2019, 6, 095328.

[120] Kowalski G., Kijowska K., Witczak M., Kuterasiński Ł., Łukasiewicz M. Synthesis and effect of structure on swelling properties of hydrogels based on high methylated pectin and acrylic polymers. Polymers 2019, 11(1), 114.
[121] Didehban K.H., Mirshokraie S.A., Azimvan J. Safranin-O dye removal from aqueous solution using superabsorbent lignin nanoparticle/polyacrylic acid hydrogel. Eurasian J Anal Chem 2018, 13(3), em29.
[122] Guan H.L., Yong D.L., Fan M.X., Yu X.L., Wang Z., Liu J.J., Li J.B. Sodium humate modified superabsorbent resin with higher salt-tolerating and moisture-resisting capacities. J Appl Polym Sci 2018, 135(48), 46892.
[123] Dabbaghi A., Kabiri K., Ramazani A., Zohuriaan-Mehr M.J., Jahandideh A. Synthesis of bio-based internal and external cross-linkers based on tannic acid for preparation of antibacterial superabsorbents. Polym Adv Technol 2019, 30, 2894–2905.
[124] Álvarez-Castillo E., Bengoechea C., Rodríguez N., Guerrero A. Development of green superabsorbent polymers from a by-product of the meat industry. J Clean Prod 2019, 223, 651–661.
[125] Capezza A.J., Glad D., Özeren H.D., Newson W.R., Olsson R.T., Johansson E., Hedenqvist M.S. Novel sustainable superabsorbents: a one-pot method for functionalization of side-stream potato proteins. ACS Sustainable Chem Eng 2019, 7, 17845–17854.

Andrada Serafim, Filis Curti, Elena Olăret, Carmen Nicolae,
Izabela-Cristina Stancu

2 Superabsorbent polymers for nanomaterials

Abstract: The present chapter will focus on the use of superabsorbent polymers in obtaining nanomaterials. Two main classes of products are identified: composites containing a nanophase and a superabsorbent matrix (superabsorbent polymer nanocomposites - SAPNCs) and superabsorbent nanometric formulations (superabsorbent nanogels - SAPNGs).

In SAPNC systems, the nanophase may act as filler, additional enhancer of water retention or thermal stability, or platform for controlled release of active agents (drugs, fertilizers, etc). Several factors must be considered when designing such complex materials, among which the nature of both matrix and nanophase, their combination method and interfacial interactions. The first part of this chapter discusses the above mentioned factors, offering relevant examples for various systems and preparation particularities.

The second part of the chapter is dedicated to the preparative techniques of SAPNGs, with the aim of combining the characteristics of hydrogels (e.g, internal structure and stimuli responsiveness) with the advantages derived from their nano-dimension (e.g, high surface-to-volume ratio), resulting in promising nanoplatforms that can be engineered for various applications, such as drug delivery systems or bioimaging.

Regardless of the chosen approach (either SAPNCs or SAPNGs), designing superabsorbent nanomaterials relies on merging high water retention ability with the specific features derived from the high surface-to-volume ratio of nanosized materials. This new generation of superabsorbent materials has the potential of being used in a wide range of fields including agriculture, waste management or medical.

2.1 Introduction to superabsorbent polymers for nanomaterials

Superabsorbent polymers (SAPs) and SAP nanocomposites (SAPNCs) define a fast-advancing branch in the field of polymer science with their capacity of managing high fluid absorption (averaging around 1,000 times the weight of the matrix) being of paramount importance for different applications including agriculture [1–4], hygienic products [5, 6], water purification treatments [7, 8], pharmaceuticals [9–11], and constructions (in mortar and concrete fabrication) [12–15]. The continuous progress in nanotechnology additionally adjusted the properties of SAPs, enabling the development of advanced SAP-based nanomaterials.

Since the term "nanomaterials" refers to materials with size features in the nanometer range, *SAPNCs* and *superabsorbent nanogels* (*SAPNGs*) may be considered as main classes of the family of *SAPs for nanomaterials*, and this chapter focuses on them.

https://doi.org/10.1515/9781501519116-002

When engineering predefined high absorption capacity materials tailored for specific applications, the relationship between their structure and properties is a key parameter. In general, the superabsorbent behavior of SAPs is dependent on physicochemical features of the matrix, such as (i) hydrophilicity – involving chemical functionality including hydroxyl, carboxyl, and amine; (ii) polyelectrolyte nature; (iii) water insolubility due to network-like structures formed through cross-linking; (iv) cross-linking degree; and (v) molecular weight of both macromolecules and cross-linker. In addition to solid content, the degree of cross-linking, cross-linking spacer length, and chemical nature are also main factors involved in adjusting the water absorption capacity. The entrapment of water and aqueous solutions in such materials has to be fast, with swelling equilibrium reached within minutes, and has to enable the absorbed liquid that is difficult to be released from the matrix even under mechanical stress. Regarding SAPNC materials, the properties of the nanophase interfacing with the SAP matrix are of utmost importance as well.

While it became more common to enlarge the application fields of polymers by developing composite materials, in addition to their utility as such, SAPs can be employed in the fabrication of SAPNCs and SAPNGs. Engineering such materials may provide controlled superabsorbent behavior while additionally improving other functional properties for a specific application. Three main strategies to prepare such materials are overviewed in Figure 2.1.

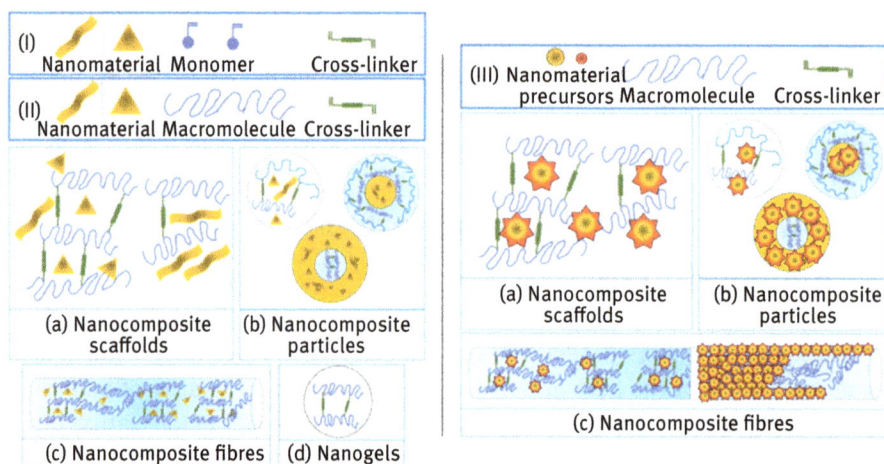

Figure 2.1: Routes to prepare SAPNCs and SAPNGs. I – starting from monomers, through network-forming (co)polymerization in the presence of nanomaterials; II – starting from macromolecules, through cross-linking in the presence of nanomaterials; and III – starting from macromolecules, nanomaterial precursors, and cross-linking systems. Main classes of products: (a) nanocomposite scaffolds, (b) nanocomposite particles with various microstructures of phase separation, (c) nanocomposite fibrous scaffolds with homogeneous composition (left panel) or with phase separation including core–shell structures (right panel), and (d) nanogels.

The first approach is to directly disperse a nanofiller in SAP precursors, with network generation through polymerization of monomers in the presence of polymerizable cross-linking agents (route I in Figure 2.1) or through cross-linking of superabsorbent macromolecules using polyfunctional reagents (route II in Figure 2.1) in homogeneous or heterogeneous reaction mixtures. Sometimes, combinatorial chemistries are applied to achieve better properties. Different polymerization techniques may be used, such as bulk polymerization, polymerization in the presence of water, solution polymerization or cross-linking, suspension or inverse suspension polymerization, precipitation polymerization generating solid scaffolds, hydrated scaffolds, (nano)particles, nanofibers and microfibers, and even three-dimensional printed products. A different strategy to generate SAPNCs (route III in Figure 2.1) refers to the use of nanomaterial precursors embedded into mixtures/solutions to be used for SAP generation. Depending on the specific steps of each strategy, at one moment the precursors will generate in situ the desired nanophase. System-specific spatial homogeneity will be obtained, with nanophase distributed homogeneously or generating heterogeneous or gradient structures (Figure 2.1, route III).

2.2 Building nanomaterials with SAP matrix

2.2.1 Superabsorbent polymer-based nanocomposites

SAPNCs emerged as a new class of materials to expand the area of applications and resolve some of SAP drawbacks, such as the lack of sustainability or degradability of synthetic ones or, for example, the insufficient mechanical tunability of natural ones. Such nanostructured materials typically integrate a natural or synthetic SAP matrix interfacing with a nanomaterial of natural or nature derived. The typical role of nanomaterials may be that of reinforcing phase, additional enhancer of water retention, enhancer of thermal stability, or nanoplatform for controlled enhanced release of therapeutic agents or fertilizers. Figure 2.2 represents the improvement of water affinity by the incorporation of a hydrophilic nanofiller into the SAP. Moreover, the selection of the constituents, both SAP matrix and nanophase, as well as their combination method and interfacial interactions are critical for the properties of the resulting nanocomposite materials. For instance, the presence of clay nanostructures improved the fire resistance of polymer and the mechanical stability while the exfoliated structure of clays provided a more porous internal structured SAPNCs, which might ensure a better aeration of the soil [16]. Furthermore, zero-dimensional diamond nanoparticles were used in combination with alginate to prepare nanocomposites with potential for sunlight-responsive release systems with agricultural use [17].

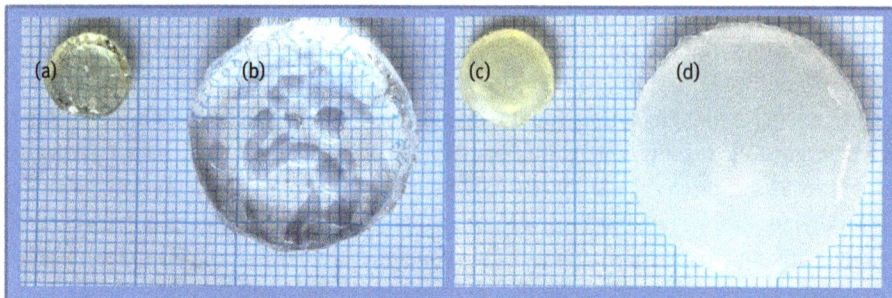

Figure 2.2: Dispersion of hydrophilic nanophase to poly(acrylamide) SAP matrix increases the water affinity of the resulting SAPNC: (a) dried SAP sample, (b) equilibrium swollen SAP sample, (c) dried SAPNC, and (d) equilibrium-swollen SAPNC.

2.2.1.1 Starting materials for SAPNCs

SAPs are the starting materials for the nanocomposite matrix, which are chosen from three main categories, namely (1) natural or nature derived, (2) synthetic, and (3) hybrid/semisynthetic polymers. Fundamental aspects regarding the polymer nature, methods of synthesis, properties, swelling mechanisms, and uses of SAPs and SAPNCs have been comprehensively reviewed [18–20]. An overview of literature reviews on SAPs is given in reference [21].

The chemistry of synthetic polymers provides a huge library of compounds and methods to obtain a diversity of controlled properties of the macromolecular matrix. The most widely used macromolecules for SAP matrices include polyacrylamide, polyacrylic acid (PAA), sodium polyacrylate (PNaA), poly(vinyl alcohol) (PVA), poly(*N*-isopropylacrylamide) (PNIPAm), polyvinylpyrrolidone (PVP), or polyvinylimidazole. The main preparation routes involve the (co)polymerization of corresponding monomers (e.g., acrylamide (AAm) and acrylic acid (AA)) in the presence of cross-linking agents (e.g., *N,N'*-methylene bisacryamide (MBA)), macromolecules, and dispersed nanophase [22–25].

Natural superabsorbent hydrogels typically include polysaccharides and proteins. Polysaccharides are appealing macromolecular building blocks of SAPNC materials considering their ability to form gels with high water content. Starch (St), sodium alginate (NaAlg), cellulose, guar gum (GG), chitosan (CS), and their derivatives are some of the commonly used biopolymers for such applications. These natural macromolecules may be used as such or modified for improved water affinity. Polysaccharide resins include oxidized, carboxymethylated, phosphated, sulfated, and grafted macromolecules. Furthermore, proteins can also be used as such or modified, and may be of various origin (extracted from different species or from different tissues). It should be stated here that for improved properties, in some situations, natural SAPs are used in combination with synthetic counterparts generating hybrid natural–synthetic SAP hydrogels. Tables 2.1 and 2.2 overview examples of SAPNCs prepared with synthetic SAP matrices (Table 2.1) and natural and hybrid matrices (Table 2.2).

Starting materials for the nanostructuring component. The nanophase used to build SAPNCs may be organic or inorganic. It should be mentioned that SAPNCs are not always referred to using the term nanocomposite, especially when the nanomaterial consists of nanostructured polymers such as cellulose nanofibrils, micellar proteins (e.g., casein), or nanosized hyperbranched polymers (e.g., polyamidoamine dendrimers). Such materials might be further combined with other nanospecies leading to new materials with attractive properties for a wide range of applications. The nature and intensity of the interactions between the nanophase and the macromolecular matrix play a decisive role in the overall behavior of the resulting SANPCs. Some representative examples are overviewed in Tables 2.1 and 2.2.

The morphology of the nanophase elemental constituents is also very important for the final properties of the SAPNC-engineered products. The shape is an additional key characteristic that directly impacts the surface-to-volume ratio playing a critical role in the tendency of nanoparticle agglomeration.

Smart SAPNCs may be built with SAP matrix responsive to external or environmental stimuli, such as temperature, pressure, moisture, pH, ionic strength, or light. These properties, corroborated with their high water absorbency, recommend them for a variety of applications, ranging from wastewater management [26–28] to controlled drug release [9, 11, 29]. Some of the most used smart SAPNCs are pH responsive, possessing reactive groups such as carboxylic, amine, or sulfonic that allow the formation of ions at particular pH values, causing them to attract or repel each other. These functional groups, responsible for the pH-active behavior, can belong either to the polymeric matrix or to the nanofiller. For example, the addition of Fe_3O_4 nanoparticles in a starch-polyacrylic acid (St-PAA) system increased the material adsorption of heavy metals and dye, but the efficiency of the removal of both metals and anionic and cationic dyes is strongly influenced by the charges of the nanocomposites [30]. pH sensitivity can be explored in drug delivery systems in cancer therapy or controlled release of active principles for the gastrointestinal tract [11]. Temperature-responsive polymeric matrices, both natural and synthetic, are also employed in the synthesis of SAPNCs. In this respect, gelatin (Gel) [31–33], PNaA [34], and PNIPAm [35, 36] homo- or copolymers may be used.

2.2.1.2 Nanocomposite preparation methods

The polymer matrix may be obtained by (co)polymerization of monomers, crosslinking of macromolecules, or combinations of the both. For example, cellulose nanocrystal-g-PAA-based nanocomposites were obtained through the free-

Table 2.1: Representative SAPNCs with synthetic SAP matrix and in situ (yellow) or dispersed (blue) nanofillers.

SAPNC	Preparative aspects	Effects, remarks|reference
Metallic nanoparticles		
Silver nanoparticles (AgNPs)		
PNaA/AgNPs	AgNPs in situ generated through the reduction of AgNO$_3$ by NaBH$_4$ followed by monomer polymerization in the presence of MBA, ammonium persulfate (APS), and tetramethylethylenediamine (TEMED) as an activator	The nanocomposite reached the equilibrium swollen state in less than 10 s; an increase in AgNP loading leads to a decrease in swelling capacity; at low amounts (20 mg/g) of AgNPs, the nanocomposite exhibits higher swelling degree when compared with PNaA; mechanical properties were improved by AgNPs. Disinfection efficiency against *Escherichia coli* and *Bacillus subtilis* – a 5.4–7.0 log reduction of viable bacteria in squeezed water after 15 s of swelling in bacterial suspension [39]
Oxides		
Iron oxide NPs (Fe$_3$O$_4$)		
P(AAm-co-sodium acrylate)/Fe$_3$O$_4$ P(AAm-co-NaA)/Fe$_3$O$_4$	Fe$_3$O$_4$ were dispersed into the monomer solution followed by polymerization in the presence of potassium persulfate (KPS), MBA, and TEMED	pH sensitive. The adsorption capacity of methylene blue increases by increasing the nanogel concentration and decreases with an increase in the ionic strength [22]
Inorganic clays		
Montmorillonite (MMT)		
P(AA-MAA)/MMT	Chains of p(AA–MAA) intercalated into the interlayers of MMT; exfoliative nanocomposite with randomly distributed separated independent layers	Microporous nanocomposite, exfoliated MMT [40]

Laponite (Lap)

PMMA/LapRDS	Polymerization of MAA in the presence of LapRDS, KPS, and MBA	Addition of LapRDS increases the pore size. High swelling degree was recorded for nanocomposites regardless of the amount of LapRDS. However for more than 5%, Lap swelling capacity decreases as it acts as a cross-linker. pH sensitive [23]

Bentonite (BT)

PNaA/MAA-modified BT	Copolymerization of sodium acrylate (NaA) and methyl acrylate groups from modified BT in the presence of APS and MBA	MAA-modified BT is homogenously dispersed within the SPA matrix, mainly as exfoliated structure. The maximum water absorbency (1,287 g/g) was recorder for 20 wt% MAA-modified BT loading. For higher loadings, a decrease in water absorbency is observed. MAA-modified BT also improved the swelling rate, water retention capacity, and thermal stability [38]

Mica (MI)

PNaA/MI	Inverse suspension polymerization of NaA cross-linked with MBA in the presence of two types of mica (K^+−MI and intercalated MI (IMI))	Water absorbency decreases with increasing MI amount. MI is exfoliated and dispersed at nanoscale, increasing particle surface area and enhancing ionic interaction and H-bonding between polymer matrix and dispersed MI phase. This increases the gel strength. The water absorbency for K^+−MI series gels was higher than trimethylammonium chloride−MI series gels [41]
Poly(AAm-co-2-acrylamido-2-methyl-1-propanesulfonic acid) P(AAm-co-AMPS)/MI	MI was dispersed in the monomer solution followed by polymerization in the presence of APS initiator, TEMED coinitiator, and MBA cross-linker	Maximum water absorbency in deionized water was obtained for P(AAm-co-AMPS-Na^+). Addition of MI decreases the water absorbency and enhances thermal stability (acting as a heat barrier). The nanocomposites assure protection by slowing down ignition time [24]

(continued)

Table 2.1 (continued)

SAPNC	Preparative aspects	Effects, remarks|reference
Multi-nanophase		
Organo-bentonite (OBT) and Fe_3O_4		
PNaA/OBT–Fe_3O_4	Copolymerization of AA (partially neutralized) in the presence of OBT–Fe_3O_4, MBA, and APS	Increasing OBT–Fe_3O_4 content, thermal stability increases. Small amounts of OBT–Fe_3O_4 (up to 3.6 g) improve swelling capacity of OBT–Fe_3O_4 PNaA. Efficient for removal of Th(IV) from water. Sensitive to pH and ionic strength [42]
Organic fillers		
Cellulose nanofibrils (CNF)		
PAA/CNF	Copolymerization of neutralized AA in the presence of CNF, urea, KPS, and MBA	The addition and increasing amount of both urea and CNF lead to a decrease in water absorption. Absorbency under load was found to be maximum when urea:AA ratio and CNF amounts were 3:10 and 5 wt%, respectively [43]
Graphene oxide (GO)		
Poly(2-acrylamido-2-methyl-1-propanesulfonic acid)/GO (PAMPS/GO)	GO was added to the monomer solution followed by polymerization in the presence of MBA and KPS	Swelling and deswelling mechanisms were best at 0.2 wt% GO; the vapor permeation increases with increasing the GO content; tensile strength is increased when added up to 0.2 wt% GO [44]

PAAm//GO	GO is dispersed into the monomer solution followed by its polymerization in the presence of MBA and APS	Thermal stability is significantly enhanced by the addition of GO; swelling behavior was investigated in distilled water, NaCl, and MgCl$_2$ solutions. Water absorbance was doubled by the addition of GO, and salt tolerance and water retention were enhanced [45]
Multiwalled carbon nanotubes (MWCNTs)		
Polyamide 6/MWCNTs (PA6/MWCNTs)	MWCNTs were dispersed in melted PA6	Hygrothermal aging weaken all mechanical properties of both PA6 and PA6/MWCNTs, but the presence of MWCNT enhances the mechanical properties of PA6 so much that it compensates the effect of hygrothermal aging [46]
p(AA-co-itaconic acid)/MWCNTs (p(AA-co-IA)/MWCNTs)	Graft copolymerization under ultrasound-assisted condition of AAm and IA in the presence of MWCNTs using APS and MBA	Thermal stability is improved in p(AAm-co-IA)/MWCNTs; swelling capacity is decreased due to the cross-linker effect of MWCNTs in hydrogels. Water retention is enhanced by nanostructures' presence. pH sensitive. Efficient for Pb (II) adsorption [47]
Carbon nanotubes (CNT)		
Polyurethane–CNT–polydopamine–octadecylamine (PU–CNT–PDA–ODA)	PU sponge etched in CrO$_3$/H$_2$SO$_4$ and CNT–PDA modified were mixed in a dopamine solution followed by the reaction with ODA	Superhydrophilic and superoleophilic composites due to PDA and ODA; CNT–PDA efficiently anchored to PU improved its thermal stability and mechanical strength. Absorption capacity increased from 22 to 34.9 times of its own weight for different oils. Reusable up to 150 times [48]

radical polymerization of AA in the presence of cellulose nanocrystals and phosphorescent Eu^{2+}/Dy^{3+}-doped $SrAl_2O_4$. The superabsorbent character was proved by uptaking more than 300 g/g water and 30 g/g saline solution. In addition, the synthesized SAPNC showed the ability to retain moisture even in extreme conditions (80 °C) for more than 3 h [37].

The nanocomposite preparation may have different degrees of complexity. It may involve the incorporation of nanofillers into the polymer matrix mainly using physical dispersion but also through in situ synthesis (as suggested in Figure 2.1). For example, silver nanoparticles are typically obtained in situ, through the reduction of $AgNO_3$ by $NaBH_4$ during the preparation of the SAP matrix.

The selection of the appropriate methods and their combination represents key elements of the synthesis strategy and decides the performances of the resulting nanocomposite materials.

Molecular self-assembly is also an important process that has to be considered, especially when biopolymers such as proteins are used. Alternating positively and negatively charged sequences may generate nanodomains such as nanofibrils within hydrogels.

Physical interactions and associations between the constituents of the nanocomposites are essential as well. The functionality of the polymer building blocks and of the nanophase has to be carefully analyzed prior to selecting the preparation method of the nanocomposite material since it can provide synthesis routes unavailable for the individual components.

Complex situations are also possible with combined mechanisms and preparative steps, in correlation with the desired application. The effect of experimental steps and order of combining starting materials is often neglected or insufficiently explained, while it may provide many opportunities to obtain new composites with finely tuned performances.

2.2.1.3 Interfacial interactions between the SAP matrix and the nanophase

The compatibility and the physicochemical interactions between their components represent important factors influencing the properties of SAP nanocomposites, in addition to the composition and individual characteristics of both polymer and nanophase. The very specific relationship nanophase–polymer application should be considered for the best possible outcome in terms of improvement of properties. Different characteristics are important to control such dependency. The interfacial interactions between the macromolecular resin and the nanofiller, and accordingly the chemical composition of the nanofiller surface, are recognized as critical for both the polymer-to-filler compatibility and nanoparticle-to-nanoparticle interactions, further deciding the dispersibility and adhesion with the SAP continuous phase.

The nature, functional groups, and presence of positive or negative charges are characteristics of each constituent in a nanocomposite. The composition, way of combining the building blocks and their interactions, further decides the identity of the resulting nanomaterial. Interfacial interactions between different constituents play a key role in describing the properties of the SAPNCs. Various physical or chemical interactions are possible, often occurring simultaneously or in complex competition. Hydrogen bonds between the matrix and carboxylate-ended nanoparticles, electrostatic interactions between base and acidic polymers and nanofillers, and ionic gelation of macromolecules at the surface of the nanophase are only few examples. It is widely known that alginates can gel fast due to ionic cross-linking. Therefore, alginate-based SAP nanocomposites reinforced with calcium carbonate nanofillers may be obtained by dispersion of the nanomaterial into the polymer solution with ionic cross-linking of the macromolecules at the interface with the nanophase.

Furthermore, nanofillers may be pretreated with monomers or other compounds for a better dispersion and predefined interactions with the polymer matrix.

The interactions between SAP and clay nanofillers are among the most interesting, and they can be controlled through the nature of the macromolecular matrix and the fabrication method. The exfoliated configuration corresponds to the separation of the nanoclay platelets followed by their individual dispersion in the macromolecular matrix. Such configuration intensifies SAP–nanofiller interactions [22]. Laponite, for example, is a nanoclay used to prepare nanocomposites with exfoliated nanophase in a poly(methacrylic acid) matrix. Interestingly, SAPNCs can be obtained only at low concentration of the nanophase, with a maximum increase of the water retention (around 91 ± 2 g/g) at concentration of only 5% nanoclay [22]. At such a ratio between the matrix and the nanomaterial, the cations from laponite are dispersed in the polymeric network enhancing the overall hydrophilic character and the gradient of osmotic pressure. Electrostatic repulsions between the anionic groups of the SAP matrix and the negatively charged nanoplatelets are responsible for higher water uptake. Very importantly, the nanophase/SAP ratio is critical, beyond 5% of the nanofiller behaving as a physical cross-linker [22].

Bentonite (BT) is another nanomaterial with exfoliating nanostructure that can play multiple roles when modified with methacrylic acid and used to generate a superabsorbent nanocomposite with a PNaA matrix [38]. In this case, the BT plays multiple roles, from cross-linker for the polymer to cost reduction and reinforcing agent [38]. The resulting SAPNC achieved a water absorbency of 1,287 g/g after only 45 min in distilled water. The clay addition to the synthetic base improved its thermal stability as well as the water uptake capacity (from 660 g/g for the pristine PNaA to 1,280 g/g for the nanocomposite). In saline solution, the water absorbency was decreased by the ionic strength when compared with distilled water, reaching only 200 g/g [38]. The authors emphasized the important effect of the initiator concentration on the water uptake ability of the resulting SAPNC. An increased dose of initiator determined a swift polymerization process, which leads to nonhomogenous cross-linking points and further to a decreased water absorbency [38].

2.2.1.4 Effect of nanophase over the SAP matrix

The addition of a nanofiller in a polymeric matrix is expected to lead to modification of the resulting material, not only in terms of mechanical properties but also with respect to water or aqueous solutions uptake, swelling and/or degradation rate, ability to generate porous structures, and so on. In the case of functionalized nanofillers, these influences are even more severe, since the interactions between the functional groups on the nanophase surface and the groups of the polymeric matric may lead to significant changes in the overall behavior of the synthesized materials. For example, St-based SAPNCs reinforced with clinoptilolite (clino) were prepared using one-pot synthesis of a semi-IPN formed of clino-filled AA–AAm-grafted St and PVA [49]. The finite element-scanning electron microscopy micrographs registered for both the composite and the neat hydrogel showed that the addition of clino leads to more porous scaffolds with a higher contact surface area. An in-depth characterization of the synthesized materials was performed regarding water absorbency and swelling kinetics, swelling capacity at various pH values and in NaCl solution, water absorbency under load, and water retention behavior. The study revealed that, when compared with the neat hydrogel, the superabsorbent nanocomposite exhibited a higher swelling capacity and reached the swelling equilibrium faster. The authors attributed this behavior to (1) the higher contact surface area of the nanocomposites and (2) the electrostatic repulsive forces between the hydroxyl groups on the polymeric backbone and the negative charges on clino's surface, leading to a more expanded network. Similar behavior was exhibited when the swelling capacity was investigated under load. The tests regarding the absorbency potential at various pH values and in saline solutions showed that the presence of clino improves the swelling ability of the materials, higher values being registered for the St-g-p(AA-co-AAm)/PVA/clino when compared with the St-g-p(AA-co-AAm)/PVA samples for all studied pH values and NaCl concentrations, respectively. The presence of clino was also considered to be responsible for the higher water retention capacity exhibited by the SAPNCs due to the strong hydrogen bonding between the absorbed water and the nanospecies. Similar results in terms of water absorbency were also registered for other systems, such as St-g-poly(AAm-AA)/SiO$_2$ and St-g-poly(AAm-AA)/Ca^{2+} [50], St-g-poly(AM–AA)/MMT [51], and CS-g-poly(AA)/MMT [52]. Moreover, several studies showed that higher water absorbency was registered when the materials were incubated in monovalent salt solution when compared with polyvalent salt solutions due to the formation of intramolecular complexes between the carboxylate groups and the multivalent cations present in the incubation media [51, 53].

Swelling–deswelling tests in a preestablished pH range are also often performed when characterizing SAPNCs, a good reusability recommending them for applications such as drug release [54]. Several systems showed remarkable pH reversibility [54, 55] behavior that was attributed to the supplementary cross-linking, which may be established between the functional groups of the nanofiller and the polymeric matrix. Other effects of the nanofiller addition on the SAP phase are overviewed in Tables 2.1 and 2.2.

2.2.2 SAPs for preparation of SAPNGs

Although considerable research has been focused on macroscopic SAP hydrogels, the development of their structures with micro- and nanoscale dimensions gained increasing attention in the last decades. SAPNGs were proposed as innovative and versatile structures in various areas of polymer chemistry and physics, material science, pharmaceutical, and medical fields. SAPNGs are nanosized particles synergistically combining characteristics of hydrogels such as their internal structure and stimuli responsiveness with selected advantages derived from their nanodimension such as high surface-to-volume ratio, resulting in promising nanoplatforms that can be engineered for various applications, including drug delivery systems and bioimaging [2, 83–85]. A large variety of biomaterials, including Gel, collagen, fibrin, polypeptides, CS, NaAlg, methyl cellulose, and PNIPAm, were used for nanogel (NGs) formulations. Different preparative techniques for hydrogel synthesis were adapted for SAPNGs development: dispersion polymerization, precipitation polymerization, nanoprecipitation, miniemulsion, microemulsion, cross-linking of micelles, and micro-molding techniques [86–89]. An overview on preparation routes, stimuli sensitivity, and biomedical applications of NGs is comprehensively reviewed in reference [90]. The successful preparation of NGs depends on three key factors: the chemical composition of the polymer, the control over the particles size, and the cross-linking procedure [91].

Nanoengineering SAPNGs in comparison to other SAPs determine significant advantages, especially for drug delivery applications. Their dimension, charge, softness property, hydrophilicity, swelling degree, and degradability can be tailored by varying the chemical composition and the concentration of components [90]. The outstanding physicochemical features of NGs are related to their nanoscaled dimension, and refers mainly to their high water uptake in a controlled manner, electromobility, superior colloidal stability, considerable mechanical integrity, structural flexibility, dispersibility in biological fluids, biodegradability, and stimuli-responsive behavior [84, 89, 92].

Considering that one of the main uses of NGs is related to pharmaceutical and biomedical fields to deliver smart drug carriers, such nanosized structures do not induce negative biological responses at molecular, cellular, or organ level [91]. NGs are generally biocompatible without toxicity, aspects largely attributed to the high water content and hydrophilic nature due to the functional groups in the polymeric chains [91].

Currently, SAPNGs are involved in considerably more applications compared to their macrosize counterparts, especially as smart nanocarriers in drug and gene delivery, cancer treatment, and cholesterol controlling, as well as for enzymes encapsulation

Table 2.2: Representative natural and hybrid SAPNCs with nanofillers generated in situ (yellow), dispersed (blue), or both in situ and dispersed (green).

SAPNC	Preparative aspects	Effects, remarks|reference
	Metallic NPs	
Silver nanoparticles (AgNPs)		
GG/PAA/AgNPs	Free-radical graft copolymerization of GG and PAA in the presence of MBA and APS; in situ generated AgNPs through the reduction of $AgNO_3$ by $NaBH_4$	Swelling capacity is controlled by the composition of the hydrogel precursors; pH sensitive [29]
NaAlg/PVA-g-PAAm/AgNPs	AgNPs in situ generated through the reduction of $AgNO_3$ by $NaBH_4$. Monomers were polymerized in the presence of MBA and APS	Presence of AgNPs significantly improves thermal stability of the hydrogel and provides antibacterial activity against both gram-positive and gram-negative microorganisms. Size of AgNPs is dependent on MBA content and NaAlg/PVA weight ratio: high content of MBA produces lower size AgNPs while increasing the amount of NaAlg leads to increased size AgNPs [56]
Poly(hydroxyethylmethacrylate-co-AA)/gum arabic/AgNPs (poly(HEMA-co-AA)/GA/AgNPs)	AgNPs in situ generated by the reduction of $AgNO_3$ by trisodium citrate ($Na_3C_6H_5O_7$). Polymerization was performed in the presence of MBA (cross-linker), APS (initiator), and $CuSO_4$/glycine (catalyst)	The nanocomposite swelled faster and reached a higher swelling degree, exhibited antibacterial activity against E. coli, and improved the thermal stability [1]
Carboxymethyl cellulose-g-PAA/AgNPs (CMC-g-PAA/AgNPs)	AgNPs in situ generated through the reduction of $AgNO_3$ by glucose. MBA (cross-linker) and KPS (initiator) were used in the polymerization reaction	The swelling ratio is increased with the addition of AgNPs from 480 g/g in distilled water and 55 g/g in 0.9 wt% NaCl solution to 860 and 74 g/g, respectively (50 mg $AgNO_3$ added during synthesis). Antibacterial activity against Staphylococcus aureus and E. coli was obtained. Thermal stability was improved [57]

PAAm–soy protein on AgNPs (PAAm–SP@AgNPs)	PAAm–SP is immersed in a solution of AgNO$_3$ followed by the reduction of AgNPs in the presence of trisodium citrate	PAAm–SP@AgNPs' swelling is higher than PAAm–SPs. Swelling is higher in basic media than in acidic media where protonation of hydrophilic moieties leads to intermolecular H-bonding hindering diffusion of water. The antibacterial activity increases with AgNPs content. Ciprofloxacin release was 95.27% for PAAm–SP@AgNPs and 75% for PAAm–SP after 6 h [58]
Gold nanoparticles (AuNPs)		
CS matrix–pectine shell/AuNPs (CS–PT/AuNPs)	AuNPs in situ generated through the reduction of HClAu$_4$ in polysaccharide-based solutions	The nanocomposite hydrogel is reinforced by AuNPs that increase elastic modulus (from 3.5 to 7.6 Pa) and decrease water uptake from 4,465% to 2,976% when compared with bare hydrogel [59]
Salep-g-PAA/AuNPs	AuNPs obtained through the reduction of HClAu$_4$ with sodium borohydride were dispersed in salep solution followed by monomer addition and polymerization in the presence of APS and MBA	pH-sensitive systems; thermal stability is improved by AuNPs presence; the nanocomposite system is suitable for controlled drug release in acidic environments [11]
Oxides		
Clinoptilolite (clino)		
St-g-p(AA-co-AAm)/PVA/clino	Clino was dispersed into the hydrogel precursor, followed by polymerization of synthetic monomers in the presence of St	Higher swelling equilibrium when compared with the control hydrogel; enhanced mechanical strength; pH sensitive [49]

(continued)

Table 2.2 (continued)

SAPNC	Preparative aspects	Effects, remarks|reference
(1) St-*g*-p(AA-*co*-AAm)/clino (2) St-*g*-p(AA-*co*-AAm)/modified clino (3) St-*g*-p(AA-*co*-AAm)/SiO$_2$	AA and AAm grafted onto St through free-radical polymerization (MBA – cross-linker, APS – initiator, and sodium bicarbonate – pore forming agent). Clino/modified clino and SiO$_2$, respectively, were ultrasounded in acetone and added into the monomer mixture, followed by polymerization	Equilibrium swelling in phosphate buffered saline was lower than in distilled water due to the presence of additional cations resulting in a "charge screening effect". Blood clotting was improved by the addition of nanospecies, best values being registered by the addition of modified clino [50]

TiO$_2$ NPs

NaAlg–PAA/TiO$_2$	Graft copolymerization of AA onto NaAlg (MBA – cross-linker and KPS –initiator) in the presence of TiO$_2$ nanoparticles predispersed in deionized water	The addition of 20% nanofiller with respect to NaAlg content (w/w) increases the adsorption of cationic dye from 80% (registered in the case of the filler-free hydrogel) to 99.4%; increasing the AA content from 0.6 to 1.2 M increases the adsorption of dye from 93% to 99%; pH responsive (maximum removal of the dye was registered at pH > 4) [26]

Iron oxide NPs (IONPs)

CMC-*g*-PAA/IONPs	IONPs in situ generated in the hydrogel precursor by coprecipitation of iron precursor (FeCl$_2$·4H$_2$O and FeCl$_3$·6H$_2$O) in the presence of NH$_4$OH. Polymerization was performed with MBA as a cross-linker and APS as an initiator	The composite hydrogel possesses superparamagnetic properties. The addition of IONPs increases the overall porosity, thus enhancing the swelling capacity with maximum swelling capacity recorded for basic media [60]
St-PAA/Fe$_3$O$_4$	Fe$_3$O$_4$ suspension (obtained by conventional coprecipitation) was added to the polymeric precursor followed by polymerization of synthetic monomers in the presence of St. MBA was used as a cross-linking agent and APS as an initiator	Fe$_3$O$_4$ nanoparticles improve the kinetic adsorption process of heavy metals and dyes. After seven consecutive adsorption–desorption cycles, the adsorption capacity was found to be 89.1% of the original capacity for the St-PAA/Fe$_3$O$_4$ while for the St-PAA, it was ~66.0% [30]

Carboxymethyl St-g-PVI/PVA/Fe$_3$O$_4$ (CMSt-g-PVI/PVA/Fe$_3$O$_4$)	Fe$_3$O$_4$ nanoparticles were added to the beads precursor that was dropped into an acetone solution of boric acid followed by cross-linking with glutaraldehyde	Fe$_3$O$_4$ provided sufficient magnetic strength to the hydrogel beads and improved the thermal stability. Suitable for dye (crystal violet – cationic and congo red – anionic) and heavy metal ion (Cu(II), Pb(II), and Cd(II)) removal from wastewater [61]
Inorganic clays		
Montmorillonite (MMT)		
NaAlg-g-poly(AA-co-AAm)/MMT	MMT suspension added to the NaAlg-containing polymerization mixture, followed by the polymerization of the synthetic monomers	Addition of MMT from 5 to 15 wt% increased the water retention from 225.3 to 460.016 g/g, while further increase of MMT content from 15% to 20% decreased the water affinity; pH sensitive [62]
κ-Carrageenan–PVA/MMT (κC–PVA/MMT)	Direct mixing of the two polymers in the presence on MMT, followed by physical cross-linking of PVA (through repeated freeze-thawing cycles) and chemical cross-linking of κC through immersion of the synthesized materials in KCl solution	The presence of NaMMT considerably decreases the swelling ability of the materials, but increases the dye adsorption capacity; nonresponsive to pH changes; the presence of di- and trivalent cations in the incubation media significantly decreases the adsorption capacity of the nanocomposite systems [27]
κC-g-PAAm-co-polyitaconic acid/MMT (κC-g-PAAm-co-PIA/MMT)	Graft copolymerization of AAm and IA onto κC in the presence of MMT, MBA (cross-linker), and APS (initiator).	Swelling capacity decreases with the increase of MMT content; high pH sensitivity; lower water absorbency in di- and trivalent salt solutions when compared with monovalent cation salt solution [63]
Gel-g-(AA–AAm)/MMT	AA and AAm were grafted onto the Gel backbone in the presence of citric acid monohydrate (cross-linker) and MMT.	Enhanced water absorbance ability of the nanocomposites in distilled water and good release ability of vitamin B12 with significant dependence on the pH of the medium [31]

(continued)

Table 2.2 (continued)

SAPNC	Preparative aspects	Effects, remarks|reference
St-*g*-poly(AAm–AA)/MMT	MMT was dispersed into the reaction mixture; polymerization was performed in the presence of APS (initiator) and MBA (cross-linker)	The nanocomposites showed higher swelling ability in NaCl aqueous solution when compared to $CaCl_2$ and $FeCl_3$ aqueous solutions of the same concentrations; increasing the salt concentration leads to decreasing swelling ability [51]
CS-*g*-poly(AA)/MMT	In situ intercalative polymerization among chitosan and AA in the presence of MMT (nanofiller), MBA (cross-linker), and APS (initiator). MMT was added to the reaction mixture (one-step synthesis) or dispersed in water and subsequently added to the reaction mixture (two-steps synthesis)	The two-steps synthesis method leads to materials with lower water absorbency; all nanocomposites are pH responsive (water absorbency increasing with the increasing pH value); MMT content of up to 2% improves swelling in distilled water and 0.9% NaCl aqueous solution [52]
NaAlg-*g*-PAAPVP/MMT	Simultaneous chemical and physical processes (graft polymerization of AA monomers onto NaAlg backbone, cross-linking reaction in the presence of MBA, and interpenetration of linear PVP chains) in the presence of MMT dispersion in water	Optimal synthesis parameters were established using the statistical method of analysis of variance (ANOVA). MMT leads to a more porous structure, with greater contact area, which leads to a higher equilibrium swelling capacity but with a lower swelling rate; the pH and salt solution type and concentration of the incubation media influence the swelling behavior of the nanocomposites; the nanocomposites have improved swelling–deswelling ability when compared with the neat hydrogel [64]
NaAlg-*g*-PAMPS/MMT	Simultaneous graft polymerization of AMPS and NaAlg and cross-linking performed in the presence of MMT	Increasing the MMT loading leads to a lower swelling ability in water and multivalent cationic solutions and higher swelling in anionic salt solutions [65]

Material	Method	Findings
CS-g-vinyl-MTT	Vinyl-MTT was dispersed into the polymeric mixture followed by polymerization of AA in the presence of APS and MBA	Vinyl-MTT enhances thermal stability of CS-g-PAA. Swelling degree decreases with increasing vinyl-MTT content. Salt and pH sensitive. Antifungal and antibacterial activities against both gram-positive and gram-negative microorganisms [66]
(1) Vinyl-modified alkylated CS-poly(sodium 4-styrenesulfonate-[(acryloylamino) propyl] trimethylammonium chloride/MMT (CSAV-poly(SSNa-ClAPTA)/MMT) (2) CSAV-poly(SSNa-ClAPTA)/Hydrotalcite (CSAV-poly(SSNa-ClAPTA)/HT)	Monomers and CSAV were added to MMT dispersion/HT dispersion followed by polymerization in the presence of APS, MBA, and TEMED	No significant differences were found in terms of swelling capacity between the MMT and HT with different concentrations (0, 3, 6, and 9 wt%). The addition of 9 wt% MMT and HT increased the storage modules to approximately 82 and 67 kPa, respectively [67]

Rectorite (REC)

Material	Method	Findings
GG-PNaA/REC	Acidified REC (H^+−REC)/organified REC (CTA^+−REC) were dispersed into the precursor mixture composed of GG and NaA followed by polymerization in the presence of APS and MBA	REC was homogeneously dispersed in GG-g-PNaA matrix: exfoliated for H^+−REC and intercalated for CTA^+−REC; water absorbency is influenced by organification degree, a maximum absorption (608 g/g) for an organification degree of 9.67 wt% was reached [68]

Cloisite

Material	Method	Findings
GG-g-NaA/cloisite	Cloisite was dispersed into the GG aqueous solution through ultrasonication. After the addition of monomer (NaA), initiator (APS) and cross-linker (MBA), the polymerization was performed in a 800 W microwave reactor	Incorporation of 10% cloisite enhances the swelling ability of the nanocomposite in water, saline solution, and incubation media with various pH values (acid, neutral, and alkaline); addition of cloisite improves dye removal efficiency (89% crystal violet removal) [69]

(continued)

Table 2.2 (continued)

SAPNC	Preparative aspects	Effects, remarks|reference
Fullers' earth (FE)		
Sugarcane bagasse-chitin-PAAm/FE (SCB-chitin-PAAm/FE)	FE added to a precursor consisting of microfibrillated SCB–chitin treated with ionic liquid and loaded with a polymerization mixture, followed by the microwave polymerization of AAm and MBA; the treatment of SCB and chitin with IL enabled better dispersion in the matrix	Self-assembly of SCB and chitin was disrupted by using IL pretreatment and ultrasonication; the loosely held particles were lost after the first cycle of swelling; nanocomposite with microbial resistance; pH sensitive [70]
Bentonite (BT)		
PNaA/BT	Polymerization of NaA cross-linked with sugar in the presence of BT pretreated with hydrochloride solution of AAm	In general, the superabsorbent character decreased with the increase of BT; for a loading of BT in the range of 25–36%, the water absorbency increased; increasing the degree of neutralization leads to increasing water absorbency [71]
Attapulgite (APT)		
CMCg–NaA/APT	APT nanofibrils were dispersed into the polymerization mixture containing sodium CMC, NaA, and MBA as cross-linkers; polymerization with initiation with APS; both chemically bonded APT and physically filled APT	Remarkable pH sensitivity; water absorption increased with increasing APT content with maximum swelling at 10 wt%, followed by decrease of water affinity when APT exceeded 10% [3]
NaAlg-g-PNaA/APT	NaAlg was grafted on partially neutralized NaA; polymerization was performed in the presence of APT using APS as an initiator and MBA as a cross-linker	pH responsive, good reversibility registered following swelling–deswelling cycles in the pH range of 2–7.4 [55]

Illite

St-g-poly(AAm-AMPS)/Illite	Graft copolymerization of St, AAm, and APMS in the presence of illite micropowder. MBA was used as a cross-linker and APS as an initiator	Presence of illite improved thermal stability of the superabsorbent material. Water absorbency increases with increasing the illite content up to 7 wt% with a maximum of 1,320 g/g. Above 7 wt%, decrease in water absorbency was observed [72]

Illite/smectite mixed-layer clay (I/S)

NaAlg-g-p(AA-co-styrene)/I/S	Free-radical grafting copolymerization of NaAlg, partially neutralized AA and styrene, in the presence of I/S	Good swelling–deswelling behavior in multivalent cations solutions; anionic surfactant had less influence on water absorption than cationic surfactant; nanocomposite showed improved pH response when compared with I/S-free superabsorbent [53]

Vermiculite (VMT)

GG-g-polyNaA/CTA⁺–VMT	VMT was modified with cetyl trimethylammonium bromide (CTAB) resulting in organo-VMT (CTA⁺–VMT); solution polymerization of GG and partially neutralized NaA performed in the presence of CTA⁺–VMT, APS, and MBA	CTA⁺–VMT improved swelling, swelling rate, and gel strength when compared with VMT-filled compositions; swelling ability increased with increasing CTA⁺–VMT content in both water and saline solution up to 4.03 mass% CTA⁺ content followed by a decrease [73]
GG-g-PNaA/Mnⁿ⁺–VMT	VMT was modified with mono-, di-, and trivalent cations resulting in Mⁿ⁺–VMT; nanocomposites were obtained through simultaneous grafting and chemical cross-linking reaction of GG, NaA, and MBA, in the presence of VMT	The use of trivalent cations for VMT modification leads to nanocomposites with highest swelling ability; pH responsive; on–off behavior is preserved even after four cycles of subsequent incubation in media with pH 2 and 7.2, respectively [74]

(continued)

Table 2.2 (continued)

SAPNC	Preparative aspects	Effects, remarks|reference
Clay(palygorskite/kaolin/laterite/diatomite)		
Wheat Bran-g-PAA/clay (WB-g-PAA/clay)	The clay was dispersed in hydrogel precursors followed by polymerization of acrylic acid in the presence of WB, KPS, and MBA	WB-g-PAA/palygorskite showed exceptional loading percentage (81 wt%) in 9 mg/mL of urea aqueous solution while WB-g-PAA/kaolin recorded a lower percentage (63 wt%). Equilibrium water absorbency was found to be highest for WB-g-PAA/laterite (647 g/g) and lowest for WB-g-PAA/kaolin (437 g/g). WB-g-PAA/diatomite was best for water retention and controlled release of urea [75]
Multinanophase		
Graphitic carbon nitride (g-C₃N₄) and silver deposited titania (Ag@TiO₂) NPs		
PVA-St/g-C$_3$N$_4$/Ag@TiO$_2$	Cross-linking of PVA–St membranes with glutaraldehyde, in the presence of dispersed g-C$_3$N$_4$ and Ag@TiO$_2$ NPs	Loading with Ag@TiO$_2$ NPs increased the swelling, their charge facilitating water diffusion. Swelling was tested in NaCl and MgCl$_2$ solutions, blood, and simulated wound fluid. Increasing pH increases the swelling ability. Higher bactericidal activity with increasing NP content [76]
Graphene oxide (GO) nanosheets and nanohydroxyapatite (n-HAp)		
St-g-PAAm/GO/n-HAP	St-g-PAAm/GO/n-HAP synthesized via free-radical cross-linking copolymerization reaction in the presence of APS and MBA. Various ratios of in situ synthesized n-HAp were used using aqueous solutions of precursors (Na$_2$HPO$_4$ and CaCl$_2$) while maintaining a constant ratio of Ca:P = 1.5:1	Increasing the n-HAp content leads to decrease of the equilibrium swelling values and increase of cationic dye adsorption. The nanocomposites proved good reusability during adsorption–desorption tests (five cycles) [77]

Magnetic IONPs (MIONPs) and graphene oxide nanosheets (GO)

St-*g*-PAAm/MIONPs/GO	MIONPs were in situ generated from $FeCl_2 \cdot 4H_2O$ and $FeCl_3 \cdot 6H_2O$ precursors treated with NH_4OH; GO was dispersed into the polymeric mixture in which MIONS were in situ synthetized followed by polymerization in the presence of MBA and APS	Up to 0.1 mg/mL GO improves the Hg^{2+} adsorption due to its hydrophilic oxygenated groups. Above this concentration, the adoption decreases. Hg^{2+} adsorption capacity is increased by increasing the MOINPs amount. Morphology is significantly influenced by MIONPs presence: roughness and porosity are increased [25]

Organic fillers

Cellulose nanowhiskers (CNWs)

St-*g*-PNaA/CNWs	CNWs were dispersed into the reaction mixture. Polymerization was performed in the presence of KPS (initiator) and MBA (cross-linker)	Swelling ability and kinetics were correlated with the nanomaterials composition (NaA:St ratio, amount of MBA, and CNWs). Swelling is influenced by the pH and ionic strength of the incubation media [78]

Cellulose nanofibrils (CNF)

CS-*g*-PAA/CNF	CNF were added into reaction mixture, and polymerization was performed using KPS as an initiator and MBA as a cross-linker	Based on an analysis of variance (ANOVA), the swelling behavior is mostly influenced by the amount of cross-linker (almost 40% of contribution in the response), followed by the amount of CNF (aprox. 30% of contribution); pH responsive; good swelling–deswelling in incubation media with pH 2 and 8 [54]
CMC-*g*-p(AA-*co*-AAm)/CNF	CNF was added to the precursor composition followed by polymerization of monomers in the presence of APS and MBA	Maximum water uptake was obtained for 2.5 wt% of CNF loading (above this concentration a decrease in water uptake capacity was observed, and it was associated to an enhance of cross-linking density). pH and salt sensitive [79]

(continued)

Table 2.2 (continued)

SAPNC	Preparative aspects	Effects, remarks|reference
Graphene oxide (GO)		
Microcrystalline cellulose/GO (MCC/GO)	GO was dispersed into LiBr aqueous solution followed by dissolution of MCC and conversion to aerogels by a solvent exchange with deionized water and freeze–drying procedure	GO is found to be exfoliated and unfolded in aerogel structures; GO improves stability of aerogel in water, ethanol, and cyclohexane; absorption amount of methylene blue is significantly increased in the composite aerogel when compared with bare MCC aerogel [80]
CMC-g-PAA/GO	GO was dispersed into the polymeric mixture followed by polymerization in the presence of APS and MBA. AA with various neutralization degree was used	Water absorbency increases by increasing the content of GO (up to 0.6 wt%). Water retention and thermal stability are improved by the GO's presence [81]
Xanthan gum/PAA/GO (XG/PAA/GO)	XG, KPS (initiator), AA, water, and a synthetic acrylic-urethane cross-linking agent were added to a GO dispersion	GO improved thermal stability; GO affects the swelling capacity; pH sensitive; with the increase of GO's concentration, the absorption efficiency of methylene blue is also increased [28]
Detonation nanodiamonds (DND)		
NaAlg/DND	NaAlg was added to a DND dispersion followed by beads formation through ion gelation technique	Maximum water absorption (1.65 g/g) was reached for 0.8 mg/mL of DND; pH and sunlight sensitive; water release controlled by the amount of DND, light intensity, and time of exposure [82]

to increase the biocatalytic activity and stability [12]. Such tendency is dictated by the physical and chemical properties of SAPNGs systems that facilitate an increase of the therapeutic efficiency of the encapsulated drug and reduction of nonspecific toxicity due to the enhanced permeation and retention effect, overcoming the limitations of micron-size structures [87, 91]. The large surface area of SAPNGs is essential for enhanced bioconjugation, high water uptake, appropriate biocompatibility, and high loading efficiency of various bioactive compounds. Therefore, the versatile architecture and appealing properties of SAPNGs are those that recommend them as attractive nanocarriers to bring therapeutic cargos through various biological membranes and release them at the target site upon a biological, chemical, or physical trigger. A large variety of molecules, such as bioactive molecules (drugs, peptides, proteins, antigens, genes, and DNA) and inorganic molecules (quantum dots, AgNPs and AuNPs, and magnetic nanoparticles), can be used with SAPNGs, in comparison with their macro-sized hydrogels that have some limitations [84, 93].

The incorporation of cargo molecules into the NG network requires physiochemical interaction between the functional groups of polymeric chains and the loaded molecule [83]. This type of functional groups from the polymeric network has a remarkable influence on drug delivery and release properties [84]. The physical interactions of hydrogen bonding or van der Waals forces, involved in the bioconjugation of therapeutic agents and functional groups, facilitate the drug-carrying efficiency [84, 91]. For instance, the small drug molecules can be loaded through the combination of electrostatic and hydrophobic interactions or by hydrogen bond formation [91]. Other benefits of SAPNGs over conventional and macro-sized delivery systems include high stability for maintaining the desired dimensions for passive targeting through the enhanced permeation and retention effect, and absence of drug leakage that is avoided due to their cross-linked structure [84, 85, 87, 93]. The size control of SAPNGs during drug delivery can also influence the cytotoxicity. Their metabolization into harmless components inside the body highly depends on the preparation method. NGs can be proficiently metabolized by the target cells, avoiding the accumulation in non-desired tissues; their efficiency as nontoxic delivery vehicle is proved by the lower therapeutic dosage used and by minimizing the harmful side effects [83]. NGs varying between 10 and 100 nm can cross the biological barriers through tissue diffusion, can improve the blood circulation time, and can escape from hepatic filtration [84, 91, 92]. Such drug-loaded NGs can be injected directly to the blood stream due to their small size and can cross the blood–brain barrier while inhibiting rapid clearance mechanisms at the same time [92].

The mechanism for drug release mostly depends on the polymer type used in the SAPNGs network, applied cross-linking method, and other factors such as environmental or biological triggers [84, 91]. For a specific drug release, smart SAPNGs characterized by a stimuli responsiveness behavior were proposed. Such systems undergo adequate transitions under endogenous (redox potential, reactive oxygen species, pH, and enzymes activity) or exogenous stimuli (light, ultrasound, electricity, temperature, voltage, and magnetic fields) [87, 89, 91, 93]. The responses to such stimuli commonly cause conformational or structural differences in the SAPNG network, which dictate the on-demand triggered release of any entrapped cargo [89, 90]. For instance, volume phase transitions of SAPNGs were manifested due to the interaction with fluid molecules determining the swelling or deswelling of the cross-linked network [90]. The swelling rate depends on the structural characteristics of the SAPNGs, chemical composition, hydrophilicity, and cross-linking degree, which controls the freedom of conformational mobility of the polymeric chains [84, 86, 94].

Currently, pH-responsive SAPNGs have received significant attention due to their release mechanism of various cargos with biological relevance, highlighting their potential applications in drug delivery systems. Among SAPNGs with pH-sensitive functionalities, those based on poly (L-histidine), poly(L-lysine), poly(L-aspartic acid), and poly(L-glutamic acid) have been reported as novel smart nanocarriers [9]. Poly(L-histidine)-based SAPNGs have determined a selective drug release at lower pH, while other types of SAPNGs with potential for the treatment of multidrug resistance tumors have showed a reversible swelling and a drug release after alternating the pH [9]. In addition, various SAPNGs based on natural, synthetic, and hybrid systems are overviewed in Table 2.3.

The pH-dependent release behavior of loaded cargo can be correlated to the swelling property. SAPNGs can be designed to reach their highest swelling degree in acidic media and, subsequently, release the entrapped cargo due to major increase in the volume [91, 93]. This approach gained interest as there are significant variations in pH value inside the body fluids, pH 1.2 in the stomach or the pH 6.8 in the intestine. For example, the water solubility at physiological pH and the pH-responsive swelling of CS SAPNGs highlight their potential as attractive carriers for a large variety of biomedical applications [94]. The SAPNGs based on PNIPAm functionalized with amino-phenylboronic acid have also presented a selective release profile under physiological conditions and have released the insulin as response to fluctuating glucose level according to the pH media [90, 95].

Besides the pH-triggered swelling of SAPNGs, other external triggers, such as the ionic strength and temperature of the environment, can also influence the swelling behavior [84, 85]. SAPNGs respond faster to swelling capacity when compared to the conventional hydrogels with micro- and macrosizes, due to their large surface area that facilitates a greater possibility of fluid exchange with the environment. The swelling rate is affected by the presence of salt solutions and can be adjusted considering the preparation method, including the initiator concentration, temperature, and cross-linking density [86]. PNIPAm colloidal SAPNGs with temperature-responsive core and pH-responsive shell were developed due to their considerable and reversible volume changes in aqueous fluids upon heating between 30 and 35 °C, at pH 2 and 7 [36].

Considerable effort has been devoted to design smart delivery formulations for cancer treatment using SAPNGs for the increase of effectiveness and safety of certain anticancer drugs (see Table 2.3). The type of polymer chains in the network

Table 2.3: Representative SAPNGs based on different types of polymers.

SAP type	SAPNG preparative aspects	Remarks|reference
	Natural or naturally derived	
CS	Reverse water-in-oil microemulsion method and cross-linking by amidation reaction using polyethylene glycol (PEG) diacid and tartaric acid	Influence of cross-linking degree on the high swelling capacity; attractive as carrier for drug delivery [94]
	Reverse microemulsion method and covalent cross-linking with PEG bis(carboxymethyl) ether)	Polyoxometalate-loaded CS SAPNGs were proposed as attractive candidates for targeted delivery in tumor treatments due to their high polyoxometalate uptake. The pH-responsive swelling and release ensured the delivery of polyanions at acidic pH [96]
	Reverse microemulsion method and cross-linking with genipin	Highly monodisperse SAPNGs; pH-sensitive water uptake [97]
Chitin	Regeneration chemistry without using organic solvents	pH-responsive behavior associated with a higher swelling at acidic pH determined the release of the encapsulated drugs (acitretin and aloe emodin) used for the psoriasis treatment [89, 98]
Gel–GA aldehyde	Inverse miniemulsion technique for both components and, subsequently, the mixing and sonication of resulted emulsions	The hemocompatibility and cytocompatibility of these SAPNGs dictated their potential in drug delivery and breast cancer therapy. Curcumin (CU)-loaded SAPNGs were proposed after reaching an encapsulation efficiency of 65 ± 3% and determined the enhanced water solubility, bioavailability, and controlled release for the entrapped drug [99, 100]
Gel	Desolvation process and the UV-induced cross-linking mechanism	The particle size and thermo-responsive behavior of these UV-cross-linked Gel SAPNGs were similar to those of the NGs cross-linked with glutaraldehyde. Potential as promising colloidal drug delivery systems [32]

(continued)

Table 2.3 (continued)

SAP type	SAPNG preparative aspects	Remarks|reference
Methacryloyl fish gelatin (GelMA)	Water-in-oil nanoemulsion	The GelMA SAPNGs properties highly depend on the aqueous phase used in the emulsion, and the type of solvent used in redispersion. Potential as nanocarrier due to the particle aggregation absence, high drug loading efficiency, and cytocompatibility; Doxorubicin-loaded GelMA SAPNGs facilitated the drug release under acidic pH, highlighting the pH-dependent release behavior that would be promising for intracellular delivery [33]
NaAlg aldehyde–Gel	Inverse miniemulsion technique	Nanocarrier for CU; higher release of the loaded drug into SAPNGs in acidic conditions compared to neutral conditions; benefic influence of these SAPNGs for an enhanced water solubility and bioavailability of CU; cytotoxicity for human breast carcinoma cells [101]
CMC	Radical polymerization of methacrylated CMC with disulfide-containing cystamine bisacrylamide as a cross-linking reagent	Redox-sensitive CMC SAPNGs have demonstrated a considerable stability even in aqueous solutions with high concentration of NaCl. Compared to other drug carriers, these NGs have facilitated an outstanding drug-loading performance, possibly dictated by the increase of ionization degree and pH-sensitive swelling. A high encapsulation efficiency of 83% and a loading content of 36% for doxorubicin were achieved. These SAPNGs were internalized by the cancer cells through endocytosis determining the intracellular delivery in a reducing environment containing glutathione [102]

Lysozyme/CMC	Green self-assembly method	Favorable carriers for delivery of drugs and bioactive molecules; such methotrexate-loaded SAPNGs have generated much more significant cytotoxic effects on cancerous cells in comparison to the free drug, due to the small sizes of NGs that could easily infiltrate inside the cells, and have demonstrated an improved bioavailability and a slow release of the loaded drug [103]
CMC/casein	Self-assembly method	CU-loaded SAPNGs were coated with bilayer films containing casein and folic acid using layer-by-layer assembly. The thermal stability of coated-SAPNGs was improved compared to the raw components. Their cytotoxicity for melanoma cancer cells was enhanced due to a higher cellular uptake compared to the uncoated SAPNGs. Potential in skin cancers treatment [104]
Cholesterol-bearing pullulan	Self-assembly in water	Favorable delivery of hydrophobic drugs or proteins by their trapping inside the SAPNGs network; Prostaglandin E1 was added to the NGs to stimulate the wound healing [105]
Thiolated poly (aspartic acid) (PAS)	Oxidation of thiol-modified PAS in water-in-oil inverse miniemulsion using ultrasonication or high-pressure homogenization	Redox-responsive SAPNGs; these SAPNGs were proposed as candidates for a tumor-targeted release of encapsulated drug [85]
Synthetic		
P(AA-co-vinyl alcohol)	Hydrolyzation in alkaline media of the cross-linked poly (acrylonitrile-co-vinyl acetate) nanoparticles synthesized through soap-free emulsion polymerization	The proposed synthesis method for these SAPNGs could overcome the issues associated with the techniques in inverse conditions. pH-sensitive behavior; good colloidal stability in swollen state [106]
PNaA	Semicontinuous inverse heterophase polymerization	Temperature-sensitive PNaA SAPNGs; higher water affinity when compared to nanoparticles synthesized by inverse microemulsion polymerization [34]

(continued)

Table 2.3 (continued)

SAP type	SAPNG preparative aspects	Remarks|reference
PAA	Precipitation polymerization in an aqueous solution without using toxic surfactants and stabilizers	pH-responsive behavior; the basic pH determined a massive swelling of the SAPNGs and a high negative zeta potential [107]
PVP	Gamma radiation techniques	Potential for drug delivery applications [108]
PNIPAm	Batch inverse microemulsion polymerization	These SAPNGs were used for the development of more complex systems, such as thermoresponsive PIAAm SAPNGs/PAAm nanostructured hydrogels [35]
Hybrid	-	
PEG methyl ether-grafted Gel	Grafting of activated PEG onto Gel backbone and self-assembly method	CU was loaded in such hybrid SAPNGs, and it demonstrated superior features such as better solubility and improved chemical stability in aqueous media. The protection provided by the SAPNGs has allowed a significantly more slow and controlled release of the drug. The therapeutic efficacy of loaded CU was considerably improved compared to the free one [109]
Carboxymethyl CS and PVA	Novel in situ method without surfactant	pH-sensitive SAPNGs with improved surface property exhibited good antibacterial activity against both types of bacteria, *gram-negative E. coli* and *gram-positive S. aureus*, which increased with the increase of PVA content. The bicomponent SAPNGs have determined a water uptake up to 500% after 2 h [110]

PEG–PAS	Chemical conversion of the hydrophobic core of the cross-linked micelles into a hydrophilic PAS via alkaline hydrolysis reaction	Reduction-responsive SAPNGs for an efficient delivery and release of anticancer drugs; disintegration process was triggered by a reducing environment using thiol reducing agents whose concentrations are relatively high inside cells. The accelerated release of the drug in such conditions ensured the drug translocation to the nucleus of cancer cells [111]
PAS-graft-imidazole-PEG	Modified emulsification–evaporation method	pH-sensitive behavior (swelling and cytotoxicity); irinotecan was successfully loaded in these SAPNGs to facilitate its release dependent on the media pH. Systemic toxicity was lowered due to reversible swelling–shrinkage, remarked by pH cyclization [9]
Derivatives of PGA: diblock and triblock copolymers	Ring-opening polymerization, deprotection, chemical modification with cinnamyl alcohol and subsequent UV-mediated cross-linking	pH-responsive polypeptide SAPNGs loaded with rifampin for a smart drug delivery; the SAPNGs swelling at higher pH associated with the ionization of carboxyl groups accelerated the rifampin release. 6.4% of total drug was released at pH 4.0 and 37% at pH 7.4 [87, 112]
Poly(ethylene oxide)-b-poly(L-histidine-co-L-cystine) PGA/CS	One step ring-opening polymerization process, using high-vacuum techniques	Novel hybrid multistimuli-responsive polypeptide SAPNGs; potential as smart nanocarriers [113]
	Surface-initiated ring opening polymerization, grafting, and macromolecular cross-linking using CS	pH-responsive behavior; SAPNGs were loaded with mitoxantrone, and a high loading capacity was achieved at pH 9 compared to pH 4, highlighting the pH dependency of the drug loading capacity. The release of mitoxantrone was influenced by the environment pH. At the lowest pH, the highest drug release was obtained. The nontoxic nature of fibroblast cells and the inhibition effect of cancer cells are promising for such drug delivery systems [114]

dictates a pattern of drug release for cancer therapy, using, for example, stimuli-responsive polymers for an efficient drug release at the tumor site [91]. The release can be stimulated due to an environmental reductive stimulus in the cancer cells caused by an increased glutathione level [85].

The outstanding properties and advantages of stimuli-responsive SAPNGs have unlocked the opportunities in the field of biomedical applications and drug delivery by their promising development.

2.3 Conclusions and perspectives

The wide library of natural and synthetic SAP, the broad family of nanospecies with specific characteristics, and the different physical and chemical methods of combination hold promise for engineering SAP-based nanomaterials with controlled properties. Despite impressive evolution of SAP nanomaterials, the field is still underexplored, and it can bring added value to both SAP and nanofillers, enlarging their applications and overcoming the limitations of the individual constituents. The interest for designing nanomaterials using SAP is based on merging the high water retention capacity of the latter with the extraordinary properties derived from the high surface-to-volume ratio of the nanosized materials, leading to a new generation of superabsorbent materials to be used in a wide range of fields including agriculture, waste management, and pharmaceutical and medical uses.

References

[1] Sunny P., Ray D., Kumar P. Designing of silver nanoparticles in gum arabic based semi-IPN hydrogel. Int J Biol Macromol 2010, 46, 237–244.

[2] Bueno V.B., Catalani L.H., Daghastanli K.R.P., Cuccovia I.M., Chaimovich H. Preparation of PVP hydrogel nanoparticles using lecithin vesicles. Quim Nova 2010, 33, 2083–2087.

[3] Wang W., Wang A. Nanocomposite of carboxymethyl cellulose and attapulgite as a novel pH-sensitive superabsorbent: synthesis, characterization and properties. Carbohydr Polym 2010, 82, 83–91.

[4] Doane W.M., Doane S.W., Savich M.H. Superabsorbent polymer in agriculture applications, US 7459,501 B2, 2008.

[5] Tsigkou K., Tsafrakidou P., Zafiri C., Soto Beobide A., Kornaros M. Pretreatment of used disposable nappies: super absorbent polymer deswelling. Waste Manag 2020, 112, 20–29.

[6] Kosemund K., Schlatter H., Ochsenhirt J.L., Krause E.L., Marsman D.S., Erasala G.N. Safety evaluation of superabsorbent baby diapers. Regul Toxicol Pharmacol 2009, 53, 81–89.

[7] Lee K.R., Riley B.J., Park H.-S., Choi J.-H., Han S.Y., Hur J.-M., Peterson J.A., Zhu Z., Schreiber D.K., Kruska K., Olszta M.J. Investigation of physical and chemical properties for upgraded SAP (SiO2Al2O3P2O5) waste form to immobilize radioactive waste salt. J Nucl Mater 2019, 515, 382–391.

[8]   Wu X., Huang X., Zhu Y., Li J., Hoffmann M.R. Synthesis and application of superabsorbent polymer microspheres for rapid concentration and quantification of microbial pathogens in ambient water. Sep Purif Technol 2020, 239, 116540.
[9]   Sim T., Lim C., Cho Y.H., Lee E.S., Youn Y.S., Oh K.T. Development of pH-sensitive nanogels for cancer treatment using crosslinked poly(aspartic acid-graft-imidazole)-block-poly (ethylene glycol). J Appl Polym Sci 2018, 135, 1–10.
[10] Pourjavadi A., Barzegar S. Smart pectin-based superabsorbent hydrogel as a matrix for ibuprofen as an oral non-steroidal antiinflammatory drug delivery. Starch/Staerke 2009, 61, 173–187.
[11] Bardajee G.R., Mizani F., Hosseini S.S. pH sensitive release of doxorubicin anticancer drug from gold nanocomposite hydrogel based on poly(acrylic acid) grafted onto salep biopolymer. J Polym Res 2017, 24.
[12] Mignon A., Devisscher D., Vermeulen J., Vagenende M., Martins J., Dubruel P., De Belie N., Van Vlierberghe S. Characterization of methacrylated polysaccharides in combination with amine-based monomers for application in mortar. Carbohydr Polym 2017, 168, 173–181.
[13] Mignon A., Graulus G.-J., Snoeck D., Martins J., De Belie N., Dubruel P., Van Vlierberghe S. pH-sensitive superabsorbent polymers: a potential candidate material for self-healing concrete. J Mater Sci 2015, 50, 970–979.
[14] Mignon A., Snoeck D., Dubruel P., Van Vlierberghe S., De Belie N. Crack Mitigation in Concrete: superabsorbent Polymers as Key to Success? Mater (Basel, Switzerland) 2017, 10, 237.
[15] De Meyst L., Mannekens E., Araújo M., Snoeck D., Van Tittelboom K., Van Vlierberghe S., De Belie N. Parameter study of superabsorbent polymers (SAPs) for use in durable concrete structures. Materials (Basel) 2019, 12, 1–15.
[16] Peyravi M., Pouresmaeel-Selakjani P., Khalili S. Nanoengineering superabsorbent materials: agricultural applications. In: Prasad R. et al., eds. Nanotechnology. Springer Nature Singapore Pte Ltd, 2017.
[17] Zheng D., Bai B. Fabrication of detonation nanodiamond @ sodium alginate hydrogel beads and their performance in sunlight-triggered water release. RSC Adv 2019, 48, 27961–27972.
[18] Mignon A., De Belie N., Dubruel P., Van Vlierberghe S. Superabsorbent polymers: s review on the characteristics and applications of synthetic, polysaccharide-based, semi-synthetic and 'smart' derivatives. Eur Polym J 2019, 117, 165–178.
[19] Peyravi M., Selakjani P.P., Khalili S. Nanoengineering superabsorbent materials: Agricultural applications. In; Nanotechnol. An Agric. Paradig. 2017.
[20] Rodrigues F.H.A., Spagnol C., Pereira A.G.B., Martins A.F., Fajardo A.R., Rubira A.F., Muniz E. C. Superabsorbent hydrogel composites with a focus on hydrogels containing nanofibers or nanowhiskers of cellulose and chitin. J Appl Polym Sci 2014.
[21] Zohuriaan-Mehr M.J., Kabiri K. Superabsorbent polymer materials: a review. Iran Polym J (English Ed) 2008, 17, 451–477.
[22] Akl M.A., Atta A.M., Youssef A.E.M., Ibraheim M.A. The utility of novel superabsorbent core shell magnetic nanocomposites for efficient removal of basic dyes from aqueous solutions. J Chromatogr Sep Tech 2013, 4.
[23] Junior C.R.F., Fernandes R.S., De Moura M.R., Aouada F.A. On the preparation and physicochemical properties of pH-responsive hydrogel nanocomposite based on poly (acid methacrylic)/laponite RDS. Mater Today Commun 2020, 23, 100936.
[24] Limparyoon N., Seetapan N., Kiatkamjornwong S. Acrylamide / 2-acrylamido-2-methylpropane sulfonic acid and associated sodium salt superabsorbent copolymer nanocomposites with mica as fire retardants. Polym Degrad Stab 2011, 96, 1054–1063.

[25] Hosseinzadeh H., Ramin S. Fast and Enhanced Removal of Mercury from Aqueous Solutions by Magnetic Starch- g -Poly (Acryl Amide)/ Graphene Oxide. Polym Sci Ser B 2016, 58, 457–473.

[26] Thakur S., Pandey S., Arotiba O.A. Development of a sodium alginate-based organic/ inorganic superabsorbent composite hydrogel for adsorption of methylene blue. Carbohydr Polym 2016, 153, 34–46.

[27] Hosseinzadeh H., Zoroufi S., Mahdavinia G.R. Study on adsorption of cationic dye on novel kappa-carrageenan/poly(vinyl alcohol)/montmorillonite nanocomposite hydrogels. Polym Bull 2015, 72, 1339–1363.

[28] Hosseini S.M., Shahrousvand M., Shojaei S., Khonakdar H.A., Asefnejad A., Goodarzi V. Preparation of superabsorbent eco-friendly semi-interpenetrating network based on cross-linked poly acrylic acid/xanthan gum/ graphene oxide (PAA/XG/GO): characterization and dye removal ability. Int J Biol Macromol 2020.

[29] Singh J., Dhaliwal A.S. Synthesis, characterization and swelling behavior of silver nanoparticles containing superabsorbent based on grafted copolymer of polyacrylic acid/ Guar gum. Vacuum 2018, 157, 51–60.

[30] Saberi A., Alipour E., Sadeghi M. Superabsorbent magnetic Fe3O4-based starch-poly(acrylic acid) nanocomposite hydrogel for efficient removal of dyes and heavy metal ions from water. J Polym Res 2019, 26.

[31] Nath J., Chowdhury A., Ali I., Dolui S.K. Development of a gelatin-g-poly(acrylic acid-co-acrylamide)–montmorillonite superabsorbent hydrogels for in vitro controlled release of vitamin B 12. J Appl Polym Sci 2019, 136, 1–11.

[32] Maki Y., Saito W., Dobashi T. Preparation and thermoresponsive behaviors of UV-crosslinked gelatine nanogels. J Biorheol 2018, 32, 15–19.

[33] Kang M.G., Lee M.Y., Cha J.M., Lee J.K., Lee S.C., Kim J., Hwang Y.S., Bae H. Nanogels derived from fish gelatin: application to drug delivery system. Mar Drugs 2019, 17, 1–11.

[34] Silva J.M., Bautista F., Sánchez-Díaz J.C., Nuño-Donlucas S.M., Puig J.E., Hernández E. Synthesis of poly(Sodium Acrylate) nanogels via semicontinuous inverse heterophase polymerization. Macromol React Eng 2015, 9, 125–131.

[35] Fernandez V.V.A., Aguilar J., Soltero J.F.A., Moscoso-Sánchez F.J., Sánchez-Díaz J.C., Hernandez E. Thermoresponsive poly(N-isopropylacrylamide) nanogels/poly(acrylamide) nanostructured hydrogels. J Macromol Sci Part A Pure Appl Chem 2016, 53.

[36] Kuckling D., Vo C.D., Wohlrab S.E. Preparation of nanogels with temperature-responsive core and pH-responsive arms by photo-cross-linking. Langmuir 2002, 18, 4263–4269.

[37] Wu Y., Wang L., Qing Y., Yan N., Tian C., Huang Y. A green route to prepare fluorescent and absorbent nano-hybrid hydrogel for water detection. Sci Rep 2017, 7, 1–11.

[38] Fu L., Cao T., Lei Z., Chen H., Shi Y., Xu C. Superabsorbent nanocomposite based on methyl acrylic acid-modified bentonite and sodium polyacrylate: fabrication, structure and water uptake. Mater Des 2016, 94, 322–329.

[39] Loo S., Fane A.G., Lim T., Krantz W.B., Liang Y., Liu X., Hu X. Superabsorbent cryogels decorated with silver nanoparticles as a novel water technology for point-of-use disinfection. Environ Sci Technol 2013, 47, 9363–9371.

[40] Mu Y., Du D., Yang R., Xu Z. Preparation and performance of poly(acrylic acid–methacrylic acid)/montmorillonite microporous superabsorbent nanocomposite. Mater Lett 2015, 142, 94–96.

[41] Lee W.-F., Chen Y.-C. Effect of intercalated reactive mica on water absorbency for poly (sodium acrylate) composite superabsorbents. Eur Polym J 2005, 41, 1605–1612.

[42] Wu L., Ye Y., Liu F., Tan C., Liu H., Wang S., Wang J., Yi W., Wu W. Organo-bentonite-Fe3O4 poly(sodium acrylate) magnetic superabsorbent nanocomposite: synthesis, characterization, and Thorium(IV) adsorption. Appl Clay Sci 2013, 83–84, 405–414.

[43] Shahzamani M., Taheri S., Roghanizad A., Naseri N., Dinari M. Preparation and characterization of hydrogel nanocomposite based on nanocellulose and acrylic acid in the presence of urea. Int J Biol Macromol 2020, 147, 187–193.

[44] Gorji M., Karimi M., Mashaiekhi G., Ramazani S. Superabsorbent, breathable graphene oxide- based nanocomposite hydrogel as a dense membrane for use in protective clothing superabsorbent. Polym Plast Technol Eng 2018, 58, 1–11.

[45] Zhu Z., Sun H., Li G., Liang W., Bao X., An J., La P. Preparation of polyacrylamide / graphite oxide superabsorbent nanocomposites with salt tolerance and slow release properties. J Appl Polym Sci 2013, 129, 1–7.

[46] Engineering M., Tserpes K.I., Moutsompegka E., Murariu O., Bonnaud L., Chanteli A., Moutsompegka E., Murariu O., Bonnaud L., Chanteli A. Experimental investigation of the effect of hygrothermal aging on the mechanical performance of carbon nanotube / PA6 nanocomposite. Plast Rubber Compos 2017, 46, 1–6.

[47] Marandi G.B., Farsadrooh M., Javadian H. Synthesis of poly (acrylamide-co-itaconic acid)/ MWCNTs superabsorbent hydrogel nanocomposite by ultrasound-assisted technique : swelling behavior and Pb (II) adsorption capacity. Ultrason – Sonochemistry 2017, 49, 1–12.

[48] E.W. H. Wang L.S., Liu Y.Z.Z., Gao D., Yuan R. A novel carbon nanotubes reinforced superhydrophobic and superoleophilic polyurethane sponge for selective oil-water separation through a chemical fabrication. J Mater Chem A 2015, 1, 266–273.

[49] Olad A., Doustdar F., Gharekhani H. Starch-based semi-IPN hydrogel nanocomposite integrated with clinoptilolite: preparation and swelling kinetic study. Carbohydr Polym 2018, 200, 516–528.

[50] Mirzakhanian Z., Faghihi K., Barati A., Momeni H.R. Synthesis of superabsorbent hydrogel nanocomposites for use as hemostatic agent. Int J Polym Mater Polym Biomater 2016, 65, 779–788.

[51] Zhou M., Zhao J., Zhou L. Utilization of starch and montmorillonite for the preparation of superabsorbent nanocomposite. J Appl Polym Sci 2011, 121, 2406–2412.

[52] Zhang J., Wang L., Wang A. Preparation and properties of chitosan-g-poly(acrylic acid)/ montmorillonite superabsorbent nanocomposite via in situ intercalative polymerization. Ind Eng Chem Res 2007, 46, 2497–2502.

[53] Wang A., Wang Y., Wang W., Shi X. A superabsorbent nanocomposite based on sodium alginate and illite/smectite mixed-layer clay. J Appl Polym Sci 2013, 130, 161–167.

[54] Spagnol C., Rodrigues F.H.A., Pereira A.G.B., Fajardo A.R., Rubira A.F., Muniz E.C. Superabsorbent hydrogel composite made of cellulose nanofibrils and chitosan-graft-poly (acrylic acid). Carbohydr Polym 2012, 87, 2038–2045.

[55] Yang H., Wang W., Wang A. A pH-sensitive biopolymer-based superabsorbent nanocomposite from sodium alginate and attapulgite: synthesis, characterization, and swelling behaviors. J Dispers Sci Technol 2012, 33, 1154–1162.

[56] Ghasemzadeh H., Ghanaat F., Antimicrobial alginate / PVA silver nanocomposite hydrogel, synthesis and characterization, 21 (2014) 1–14.

[57] Shen J., Cui C., Li J., Wang L., In Situ Synthesis of a Silver-Containing Superabsorbent Polymer via a Greener Method Based on Carboxymethyl Celluloses 23 (2018).

[58] Prusty K., Biswal A., Bhusan S., Swain S.K. Synthesis of soy protein / polyacrylamide nanocomposite hydrogels for delivery of ciprofloxacin drug. Mater Chem Phys 2019, 234, 378–389.

[59] De Almeida D.A., Sabino R.M., Souza P.R., Bonafé G., Venter S.A.S., Popat K.C., Alessandro F., Monteiro J.P. Pectin capped gold nanoparticles synthesis in-situ for producing durable, cytocompatible, and superabsorbent hydrogel composites with chitosan. Int J Biol Macromol 2020, 147.

[60] Hosseinzadeh H., Javadi A., Fabrication and characterization of CMC-based magnetic superabsorbent hydrogel nanocomposites for crystal violet removal, (2016).

[61] Pour Z.S., Ghaemy M. Removal of dyes and heavy metal ions from water by magnetic hydrogel beads based on poly (vinyl imidazole). RSC Adv 2015, 64106–64118.

[62] Rashidzadeh A., Olad A. Slow-released NPK fertilizer encapsulated by NaAlg-g-poly(AA-co-AAm)/MMT superabsorbent nanocomposite. Carbohydr Polym 2014, 114, 269–278.

[63] Sadeghi M. Synthesis of a biocopolymer carrageenan-p-poly(AAm-co-IA)/montmorillonite superabsorbent hydrogel composite. Brazilian J Chem Eng 2012, 29, 295–305.

[64] Olad A., Pourkhiyabi M., Gharekhani H., Doustdar F. Semi-IPN superabsorbent nanocomposite based on sodium alginate and montmorillonite: reaction parameters and swelling characteristics. Carbohydr Polym 2018, 190, 295–306.

[65] Yadav M., Rhee K.Y. Superabsorbent nanocomposite (alginate-g-PAMPS/MMT): synthesis, characterization and swelling behavior. Carbohydr Polym 2012, 90, 165–173.

[66] Aziz M.S.A., Salama H.E. Effect of vinyl montmorillonite on the physical, responsive and antimicrobial properties of the optimized polyacrylic acid/ chitosan superabsorbent via Box-Behnken model. Int J Biol Macromol 2018, 116, 840–848.

[67] Palacio D.A., Urbano B.F. Polyelectrolyte nanocomposite hydrogels filled with cationic and anionic clays. Carbohydr Polym 2019, 115824.

[68] Wang W., Wang A. Preparation, characterization and properties of superabsorbent nanocomposites based on natural guar gum and modified rectorite. Carbohydr Polym 2009, 77, 891–897.

[69] Likhitha M., Sailaja R.R.N., Priyambika V.S., Ravibabu M.V. Microwave assisted synthesis of guar gum grafted sodium acrylate/cloisite superabsorbent nanocomposites: reaction parameters and swelling characteristics. Int J Biol Macromol 2014, 65, 500–508.

[70] Sharma M., Bajpai A. Superabsorbent nanocomposite from sugarcane bagasse, chitin and clay: synthesis, characterization and swelling behaviour. Carbohydr Polym 2018, 193, 281–288.

[71] Zhang J., Yuan K., Wang Y.-P., Gu S.-J., Zhang S.-T. Preparation and properties of polyacrylate/bentonite superabsorbent hybrid via intercalated polymerization. Mater Lett 2007, 61, 316–320.

[72] Zhou M., Zou J., Guo X., Yang Y. Superabsorbent nanocomposite and its properties. J Macromol Sci Part A 2019, 0, 1–10.

[73] Wang W., Zhang J., Wang A. Preparation and swelling properties of superabsorbent nanocomposites based on natural guar gum and organo-vermiculite. Appl Clay Sci 2009, 46, 21–26.

[74] Wang W., Zhai N., Wang A. Preparation and swelling characteristics of a superabsorbent nanocomposite based on natural guar gum and cation-modified vermiculite. J Appl Polym Sci 2011, 119, 3675–3686.

[75] Gao J., Yang Q., Ran F., Ma G., Lei Z. Applied Clay Science Preparation and properties of novel eco-friendly superabsorbent composites based on raw wheat bran and clays. Appl Clay Sci 2016, 132–133, 739–747.

[76] Ahmed A., Niazi M.B.K., Jahan Z., Ahmad T., Hussain A., Pervaiz E., Janjua H.A., Hussain Z. In-vitro and in-vivo study of superabsorbent PVA/Starch/g-C3N4/Ag@TiO2 NPs hydrogel membranes for wound dressing. Eur Polym J 2020, 130, 109650.

[77]   Hosseinzadeh H., Ramin S. Fabrication of starch-graft-poly(acrylamide)/graphene oxide/
        hydroxyapatite nanocomposite hydrogel adsorbent for removal of malachite green dye from
        aqueous solution. Int J Biol Macromol 2018, 106, 101–115.
[78]   Spagnol C., Rodrigues F.H.A., Pereira A.G.B., Fajardo A.R., Rubira A.F., Muniz E.C.
        Superabsorbent hydrogel nanocomposites based on starch-g-poly(sodium acrylate) matrix
        filled with cellulose nanowhiskers. Cellulose 2012, 19, 1225–1237.
[79]   Zhou Y., Fu S., Zhang L., Zhan H. Superabsorbent nanocomposite hydrogels made of
        carboxylated cellulose nanofibrils and CMC-g-p (AA-co-AM). Carbohydr Polym 2013, 97,
        429–435.
[80]   Wei A.X., Huang T., Yang J., Zhang N. Green synthesis of hybrid graphene oxide/
        microcrystalline cellulose aerogels and their use as superabsorbents. J Hazard Mater 2017,
        335, 28–38.
[81]   Wang Z., Ning A., Xie P., Gao G., Xie L., Li X., Song A. Synthesis and swelling behaviors of
        carboxymethyl cellulose-based superabsorbent resin hybridized with graphene oxide.
        Carbohydr Polym 2017, 157, 48–56.
[82]   Zheng D., Bai B. Fabrication of detonation nanodiamond@sodium alginate hydrogel beads
        and their performance in sunlight-triggered water release. RSC Adv 2019, 9, 27961–27972.
[83]   Sabir F., Asad M.I., Qindeel M., Afzal I., Dar M.J., Shah K.U., Zeb A., Khan G.M., Ahmed N.,
        Din F.U. Polymeric nanogels as versatile nanoplatforms for biomedical applications. J
        Nanomater 2019, 2019.
[84]   Pamfil D., Vasile C. Nanogels of natural polymers. In: Thakur V.K., ed. Springer Nature
        Singapore. 2018, 71–110.
[85]   Krisch E., Messager L., Gyarmati B., Ravaine V., Szilágyi A. Redox- and pH-Responsive
        Nanogels Based on Thiolated Poly(aspartic acid). Macromol Mater Eng 2016, 301, 260–266.
[86]   Pérez-Álvarez L., Ruiz-Rubio L., Lizundia E., Vilas-Vilela J.L. Polysaccharide-based
        superabsorbents: synthesis, properties and applications. M. M, Ed. Cellul. Superabsorbent
        Hydrogels. Polym. Polym. Compos. A Ref. Ser. Cham, Springer, 2019, 1393–1431.
[87]   Jiang Z., Chen J., Cui L., Zhuang X., Ding J., Chen X. Advances in Stimuli-Responsive
        Polypeptide Nanogels. Small Methods 2018, 2, 1700307.
[88]   Krisch E., Gyarmati B., Szilágyi A. Preparation of pH-responsive poly(Aspartic acid) nanogels
        in inverse emulsion. Period Polytech Chem Eng 2017, 61, 19–26.
[89]   Cuggino J.C., Blanco E.R.O., Gugliotta L.M., Alvarez Igarzabal C.I., Calderón M. Crossing
        biological barriers with nanogels to improve drug delivery performance. J Control Release
        2019, 307, 221–246.
[90]   Soni K.S., Desale S.S., Bronich T.K. Nanogels: an overview of properties, biomedical
        applications and obstacles to clinical translation. J Control Release 2016, 240, 109–126.
[91]   Kankala R.K., Wang S.-B., Chen A.-Z., Zhang Y.S. Self-Assembled Nanogels: From Particles to
        Scaffolds and Membranes. Elsevier Inc, 2018.
[92]   Rajput R., Narkhede J., Naik J.B. Nanogels as nanocarriers for drug delivery: a review. ADMET
        DMPK 2020, 8, 1–15.
[93]   Dannert C., Stokke B.T., Dias R.S. Nanoparticle-hydrogel composites: From molecular
        interactions to macroscopic behavior. Polymers (Basel) 2019, 11.
[94]   Pujana M.A., Pérez-Álvarez L., Iturbe L.C.C., Katime I. Water dispersible pH-responsive
        chitosan nanogels modified with biocompatible crosslinking-agents. Polymer (Guildf) 2012,
        53, 3107–3116.
[95]   Hoare T., Pelton R. Charge-switching, amphoteric glucose-responsive microgels with
        physiological swelling activity. Biomacromolecules 2008, 9, 733–740.

[96] Pérez-Álvarez L., Ruiz-Rubio L., Artetxe B., Vivanco M.D.M., Gutiérrez-Zorrilla J.M., Vilas-Vilela J.L. Chitosan nanogels as nanocarriers of polyoxometalates for breast cancer therapies. Carbohydr Polym 2019, 213, 159–167.

[97] Arteche Pujana M., Pérez-Álvarez L., Cesteros Iturbe L.C., Katime I. Biodegradable chitosan nanogels crosslinked with genipin. Carbohydr Polym 2013, 94, 836–842.

[98] Divya G., Panonnummal R., Gupta S., Jayakumar R., Sabitha M. Acitretin and aloe-emodin loaded chitin nanogel for the treatment of psoriasis. Eur J Pharm Biopharm 2016, 107, 97–109.

[99] Sarika P.R., James N.R. Preparation and characterisation of gelatin-gum arabic aldehyde nanogels via inverse miniemulsion technique. Int J Biol Macromol 2015, 76, 181–187.

[100] Sarika P.R., Nirmala R.J. Curcumin loaded gum Arabic aldehyde-gelatin nanogels for breast cancer therapy. Mater Sci Eng C 2016, 65, 331–337.

[101] Sarika P.R., James N.R., Anil Kumar P.R., Raj D.K. Preparation, characterization and biological evaluation of curcumin loaded alginate aldehyde–gelatin nanogels. Mater Sci Eng C 2016, 68, 251–257.

[102] Qian H., Wang X., Yuan K., Xie C., Wu W., Jiang X., Hu L. Delivery of doxorubicin in vitro and in vivo using bio-reductive cellulose nanogels. Biomater Sci 2014, 2, 220–232.

[103] Li Z., Xu W., Zhang C., Chen Y., Li B. Self-assembled lysozyme/carboxymethylcellulose nanogels for delivery of methotrexate. Int J Biol Macromol 2015, 75, 166–172.

[104] Priya P., Mohan Raj R., Vasanthakumar V., Raj V. Curcumin-loaded layer-by-layer folic acid and casein coated carboxymethyl cellulose/casein nanogels for treatment of skin cancer. Arab J Chem 2020, 13, 694–708.

[105] Kobayashi H., Katakura O., Morimoto N., Akiyoshi K., Kasugai S. Effects of cholesterol-bearing pullulan (CHP)-nanogels in combination with prostaglandin E1 on wound healing. J Biomed Mater Res – Part B Appl Biomater 2009, 91, 55–60.

[106] Molaei S.M., Adelnia H., Seif A.M., Nasrollah Gavgani J. Sulfonate-functionalized polyacrylonitrile-based nanoparticles; synthesis, and conversion to pH-sensitive nanogels. Colloid Polym Sci 2019, 297, 1245–1253.

[107] Mackiewicz M., Stojek Z., Karbarz M. Synthesis of cross-linked poly(acrylic acid) nanogels in an aqueous environment using precipitation polymerization: Unusually high volume change. R Soc Open Sci 2019, 6.

[108] Ges A.A., Viltres H., Borja R., Rapado M., Aguilera Y. Gamma radiation-induced synthesis and characterization of Polyvinylpyrrolidone nanogels. J Phys Conf Ser 2017, 792, 012080.

[109] Tran D.N., Nguyen T.H., Vo T.N.N., Pham L.P.T., Vo D.M.H., Nguyen C.K., Bach L.G., Nguyen D.H. Self-assembled poly(ethylene glycol) methyl ether-grafted gelatin nanogels for efficient delivery of curcumin in cancer treatment. J Appl Polym Sci 2019, 47544, 47544.

[110] Farag R.K., Mohamed R.R. Synthesis and characterization of carboxymethyl chitosan nanogels for swelling studies and antimicrobial activity. Molecules 2013, 18, 190–203.

[111] Park C.W., Yang H.M., Woo M.A., Lee K.S., Kim J.D. Completely disintegrable redox-responsive poly(amino acid) nanogels for intracellular drug delivery. J Ind Eng Chem 2017, 45, 182–188.

[112] Ding J., Zhuang X., Xiao C., Cheng Y., Zhao L., He C., Tang Z., Chen X. Preparation of photo-cross-linked pH-responsive polypeptide nanogels as potential carriers for controlled drug delivery. J Mater Chem 2011, 21, 11383–11391.

[113] Bilalis P., Varlas S., Kiafa A., Velentzas A., Stravopodis D., Iatrou H. Preparation of hybrid triple-stimuli responsive nanogels based on poly(L-histidine). J Polym Sci Part A Polym Chem 2016, 54, 1278–1288.

[114] Yan S., Sun Y., Chen A., Liu L., Zhang K., Li G., Duan Y., Yin J. Templated fabrication of pH-responsive poly(l-glutamic acid) based nanogels via surface-grafting and macromolecular crosslinking. RSC Adv 2017, 7, 14888–14901.

María Fernanda Baieli, Nicolás Urtasun

3 Biomedical applications of hydrogels in the form of nano- and microparticles

Abstract: Hydrogels are hydrophilic three-dimensional networks of polymeric chains randomly cross-linked by physical or chemical bonds that are able to adsorb large amounts of water. They can be shaped into different physical conformations such as films, tablets, sponges, fibers, or particles in the range of nano- and microscale. This last conformation that includes numerous multiparticulated forms – from particles without any form to spheres – has directly been used in different biomedical applications or as a part of another conformation. Normally, a particulate form of a hydrogel is used when the biomedical application aims at delivering, immobilizing, and/or purifying a biomolecule, or when a biological fluid adsorption capacity is necessary. Here, the role of different hydrogels in the form of particles – with different polymer nature, composition, polymerization, cross-linking, and chemical modification – will be discussed. Special attention has been given to the specific polymer characteristics that bring a desired property to the final product in the areas of drug delivery, gene carriers, tissue engineering and scaffolds, biosensors, cosmetics, and diagnostic imaging.

3.1 Introduction

Hydrogels are hydrophilic three-dimensional networks of polymeric chains randomly cross-linked by physical or chemical bonds. These systems have the ability to swell and adsorb large amount of water or biological fluids in the interstitial spaces between the chains without losing their structure. This water sorption capacity of the hydrogels especially depends on the nature and density of the used polymer, the molecular weight, the presence of functional groups ($-OH$, $-CONH$, $-CONH_2$, $-SO_3H$, etc.), the use of a cross-linker, among other factors [1–3].

The spread and development of hydrogels began around the 1960s, when Wichterle and Lim synthesized poly(2-hydroxyethyl methacrylate) (pHEMA) gels to be used in contact lenses. Since those years and especially during the last two decades, a high number of polymers (natural or synthetic) with diverse properties were developed and applied for different biomedical purposes. Nowadays, a high number of commercial products based on hydrogels are already in the market [3].

Generally, hydrogels could be classified according to different criteria (Figure 3.1). According to the original source of the polymer, hydrogels could be prepared using natural and/or synthetic polymers. In addition, according to the type of cross-linking, hydrogels could have chemical cross-linking when the polymer chains interact with each other to form permanent bonds, or physical cross-link, when physical and chemical reversible interactions are formed between the polymer chains by Van der Walls

https://doi.org/10.1515/9781501519116-003

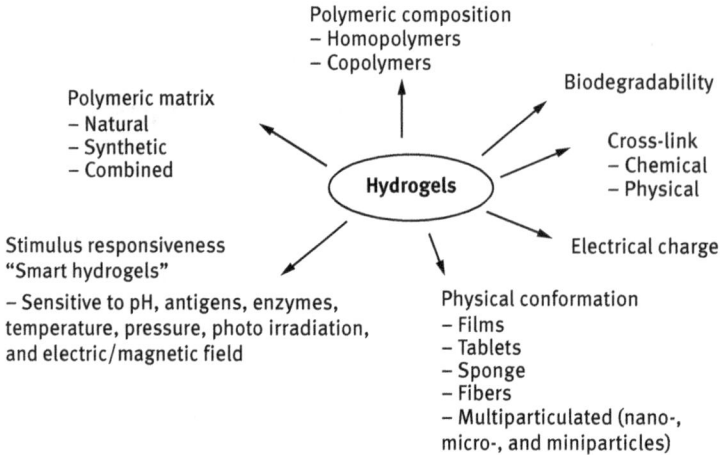

Figure 3.1: Classification of hydrogels.

forces, hydrophobic, hydrogen, and/or ionic interactions. Table 3.1 summarizes the different methods for synthesizing physical and chemical hydrogels [4].

Other types of hydrogels that have attracted attention are the "stimulus-responsive gels" or "smart hydrogels," which undergo reversible phase transitions in response to external stimuli (Figure 3.1). These gels can reversibly change their volume and/or shape in response to a slight alteration of the external stimuli, such as pH, temperature, ionic strength, light, electric or magnetic field, and biological stimulus [2]. In addition, hydrogels could be classified by their polymeric composition as homopolymers, when the polymer network derives from a single monomer, or copolymers, when two or more different monomer types were used [5].

According to the required application, a determined type of hydrogel should be developed. For example, the choice of natural polymers in hydrogels –hyaluronic acid (HA), cellulose, heparin, dextran, alginate, collagen, chitosan, and so on – for biomedical applications is advantageous due to their biocompatibility, biodegradability, and nontoxicity, whereas the choice of synthetic polymers – pHEMA, polyacrylamide (PAAm), polyvinyl alcohol (PVA), polymethyl methacrylate (PMMA), and so on – is relevant when mechanical strength and durability are required. Furthermore, the type of chemical cross-linker used usually implies a rigorous process of validation. In some cases, the cross-linker has to be absent in the final product, especially for drug delivery applications, due to its toxicity [1, 6].

Hydrogels could adopt different physical conformations, such as films, tablets, sponges, fibers, or particles, in the range of nano- and microscale. This last conformation that includes numerous multiparticulated forms, from particles without any form to spheres, has directly been used in different biomedical applications or as a part of another conformation (scaffolds, films, sponges, etc.). Generally, a particulate form of

Table 3.1: Methods for synthesizing physical and chemical hydrogels [4].

Physical methods	Chemical methods
– Warm a polymer solution to form a gel (e.g., PEO–PPO–PEO block copolymers). – Cool a polymer solution to form a gel (e.g., agarose or gelatin). – "Cross-link" a polymer in aqueous solution, using freeze–thaw cycles (e.g., PVA). – Lower pH to form an H-bonded gel between two different polymers in the same aqueous solution (e.g., PEO and PAAc). – Mix solution of a polyanion and a polycation to form complex coacervate gel (e.g., sodium alginate plus polylysine). – Gel a polyelectrolyte solution with a multivalent ion opposite charge (e.g., Na^+ alginate$^-$ + Ca^{2+} + $2Cl^-$).	– Cross-link polymers in the solid state or in solution with radiation (e.g., irradiate PEO), chemical cross-linkers (e.g., treat collagen with glutaraldehyde or a bis-epoxide), and multifunction reactive compounds (e.g., PEG + diisocyanate). – Copolymerize a monomer + cross-linker in solution (e.g., HEMA + EGDMA). – Copolymerize a monomer + a multifunctional macromer (e.g., bis-methacrylate terminated PLA + PEO + PLA + photosensitizer + visible light radiation). – Polymerize a monomer within a different solid polymer to form an interpenetrating network gel (e.g., acrylonitrile + starch). – Chemically convert a hydrophobic polymer to a hydrogel (e.g., partially hydrolyze PVAc to PVA or PAN to PAN/PAAm/PAAc).

PEO, polyethylene oxide; PPO, polypropylene oxide; PVA, polyvinyl alcohol; PAAc, polyacrylic acid; PEG, polyethylene glycol; HEMA, hydroxymethyl methacrylate; EGDMA, ethylene glycol dimethacrylate; PLA, polylactic acid; PVAc, polyvinyl acetate; PAN, polyacrylonitrile; PAAm, polyacrylamide.

a hydrogel is used when the biomedical application requires delivering, immobilizing, and/or purifying a biomolecule or when a high water – or biological fluid – adsorption capacity is necessary. The use of hydrogels in the form of particles increases the superficial contact of the fluid with the material, improves the velocity of the diffusion process when required, and adds load capacity of a specific biomolecule per gram of material.

Solute transport inside and outside the hydrogels occurs by diffusion across the water-filled regions in the space delineated by the polymer chains. Several factors could influence the movement of a molecule. Such factors include the size of the molecule in relation to pore size determined by the polymer chains, polymer chain mobility, and the existence of charged groups on the polymer, which may interact with the molecule. Polymer chain mobility is an important factor governing molecule movement within the hydrogel. Also, the diffusivity of the molecule through cross-linked hydrogel decreases as cross-linking density increases, as the size of the molecule increases, and as the volume fraction of water inside the gel decreases [7]. In addition, the hydrophobic character of the molecule and the polymer and how they interact have an impact on the solute transport. The water filling of a hydrogel with certain hydrophobicity implies the ordering of the water molecules around the polymer chains with less randomness and better hydrogen bonding than in the bulk water. Then, the interaction of a molecule with this kind of material results in the displacement of the ordered structure of water molecules in the immediate vicinity of the polymer chains, increasing the entropy of the system.

3.2 Formatting hydrogels into particles

For the preparation of nano- and microparticles, the physical and chemical cross-linking shown in Table 3.1 could be used. However, it is necessary to develop different polymerization strategies to form the particles instead of polymer aggregates. Most methods used for the preparation of particles are carried out in emulsified systems involving two steps: preparation of an emulsified system and the formation of the particles by precipitation/gelation or by polymerization of the monomers (Figure 3.2) [8]. To create an emulsified system, a high-energy emulsification technique such as mechanical shearing is used. An oil-in-water emulsified system forms droplets of uniform size where the polymer, or the monomer, is contained. Even though agitation (stirring or shaking) is the most common method to prepare an emulsion, new processes and machines have significantly evolved in the last decade to scale up the production and to control the dispersion size of the particles [9]. Once the emulsified system is achieved, the hydrogel particles are synthesized by physical or chemical cross-linking. The size and the form of the emulsified droplets determine the final size, form, and porosity of the final hydrogel particles depending on the polymer nature, the organic solvent, the viscosity of the system, the intensity of the stir, and the relation between organic solvent:water:polymer, among other factors. The second step involves a precipitation method. In this step, emulsion solvents can be removed from the organic phase by various pathways such as solvent evaporation, fast diffusion after dilution, salting out process, or by gelation of the droplets in the emulsion (Figure 3.2a) [9, 10]. In addition, an in situ polymerization could be occurring when one or two monomers are added instead of a polymer to the emulsified system. These polymerization reactions are often initiated by thermal, photo, or radiation free-radical polymerization. In these cases, it is necessary to slow down the reaction to allow the formation of particles instead of aggregates. This is accomplished by changing the temperature, the pH, and the concentration of the monomers in the reaction process (Figure 3.2b). The main problem for the emulsion polymerization method is that some unreacted monomer or oligomer may be present in the formed particle, while this may cause toxicity for clinical uses [9].

In addition to polymerization through emulsion systems, polymers soluble in aqueous media could form particles using different techniques such as solvent/nonsolvent precipitation, ionic gelation, and self-assembling (polymers that can associate

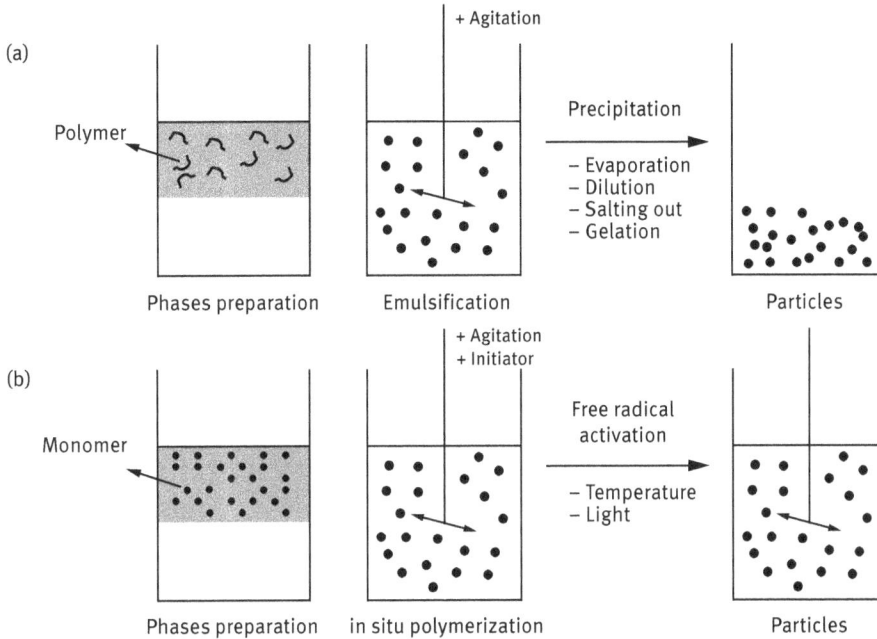

Figure 3.2: Preparation of nano- and micro-particles carried out in emulsified systems. (a) An oil-in-water emulsified system forms droplets of uniform size where the polymer is contained and after precipitation the particles are formed. (b) An oil-in-water emulsified system forms droplets of uniform size where the monomer is contained, and after a free-radical activation, the particles are formed.

together to form supramolecular nanoassemblies of spherical shapes by themselves) (Figure 3.3) [9]. These techniques are frequently selected to develop hydrogel particles from natural polymers such as alginate, chitosan, dextran, and HA – whereas emulsified systems are commonly selected for synthetic polymers such as polyethylene glycol (PEG), PVA, and pHEMA.

The next section discusses the role of different hydrogels in the form of particles – with different polymer nature, composition, polymerization, cross-linking, and chemical modification – for diverse biomedical applications where they are used (Table 3.2). Special attention will be given to the specific polymer characteristics that bring the desired property to the final product.

(a)

(b)

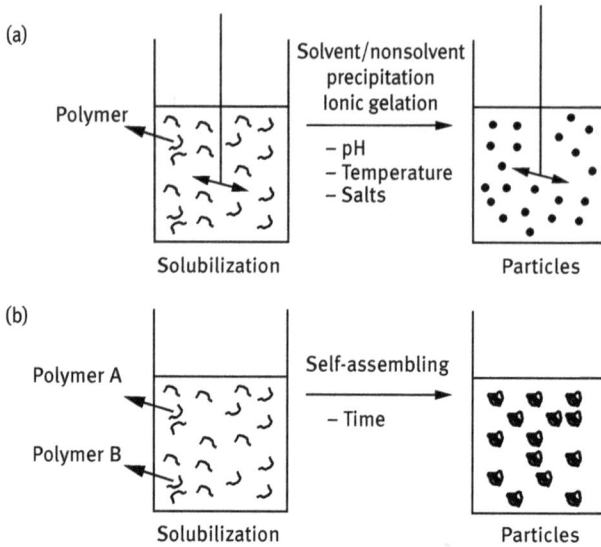

Figure 3.3: Preparation of nano- and microparticles carried out in aqueous media. (a) Formation of particles using solvent/nonsolvent or gelation precipitation. (b) Self-assembling particles – polymers that can associate together to form supramolecular nanoassemblies of spherical shapes.

Table 3.2: Summary of polymers in the form of particles most used for biomedical applications.

Application	Size	Natural polymer	Synthetic polymer	Combined polymers
Drug delivery	Nano- and microparticle	HA, chitosan, dextran, and alginate	PEG, PLGA, and PLA	Alginate–chitosan, alginate–dextran, PEG–chitosan, and PEG-PLGA
Gene carriers	Nanoparticle	Chitosan and atelocollagen	PLGA	PLGA–chitosan, PLGA–PEI, PEI–PEG, PEI–PEO, PLL–PEG and PLGA–PLL
Tissue engineering and scaffolds	Nano- and microparticle	Chitosan	PLGA and PPy	Dextran–PEG and PLGA–PEG
Biosensors using immobilized peptides and proteins	Nano- and microparticle	Dextran and chitosan	PLGA	–
Cosmetics	Microparticle	HA	–	Collagen–PMMA
Diagnostic imaging	Nano- and microparticle	Chitosan, dextran, and carboxydextran	PLL	PLA–PEG and PLGA–PEG

HA, hyaluronic acid; PLGA, polylactide-*co*-glycolide ; PLA, polylactic acid; PEI, polyethylenimine; PEO, polyethylene oxide; PLL, polylysine; PPy, polypyrrole; PMMA, polymethyl methacrylate.

3.3 Biomedical applications using nano- and micro-particles

3.3.1 Drug delivery

In drug delivery systems, pharmaceutically active compounds are loaded into a hydrogel, which is then injected into the body where the drugs are released. Including the drug into a hydrogel, instead of injecting the free drug, has the advantages to increase the bioavailability and water solubility and to reduce the antigenic activity and systemic toxicity of the loaded drug [8]. In particular, hydrogels are used to deliver their contents to a target area in the body. The hydrogel structure might protect drug from hostile environments and control drug release by changing the gel structure in response to environmental stimuli. Hydrogel's physical properties, such as swelling, surface characteristics, and mechanical strength, can be modulated by physicochemical reactions (cross-linking, thermal treatments, freeze-dried, etc.) to improve elasticity and mechanical resistance, which are important features to consider when developing delivery systems.

In general, drug load into the hydrogel takes place during the polymerization process – compounds are entrapped between polymeric chains – or by an adsorption process when the hydrogel particles are already synthesized. The kind of interactions between the drug and the polymer (Van der Walls forces, hydrophobic, hydrogen and/or ionic interactions) impacts lately in its release. According to the rate-limiting step, drug release could be diffusion controlled, which depends on the drug diffusivity across the polymeric matrix; swelling controlled, which depends on the time necessary for the solvent to penetrate inside the polymeric matrix and solubilize the drug; and chemically controlled – usually called "erosion of the matrix" – which depends on enzymatic or nonenzymatic degradation of the matrix [10].

Different materials – including natural, synthetic, and combined polymers – are used as carrier for various active agents such as antibiotics, growth factors, anti-inflammatory drugs, and proteins/peptides [11]. One of the greatest challenges that limit the success of nanoparticles in general is their ability to reach the therapeutic site at necessary doses while minimizing accumulation at undesired sites. Hydrogel nanoparticles are stable systems with the particularity of accumulating in tumors and inflammatory and infectious sites because these pathologies exhibit a nonhealthy endothelium, and nanoparticles are able to permeate and accumulate in the surrounding tissues. Furthermore, some hydrogel nanoparticles can be decorated with an antibody, antigen, protein, or integrin for the specific delivery to a particular tissue or organ [8, 12].

The desirable size range for particles used for drug delivery is between 10 and 100 nm in order to have effective body bioavailability and circulation times [13]. Their small size enables them to adhere to mucosal surfaces and, when parenterally administered, to traverse the smallest blood vessels and avoid phagocyte system. In this way, nanoparticles extend the circulation half-time of attached drugs and potentially reduce the toxic effects due to their accumulation at undesired sites [14]. Hydrogel particles can be engineered to entrap either hydrophilic or hydrophobic drug molecules, as well as macromolecules such as proteins and nucleic acids.

Polysaccharides represent the most important group of natural polymers used for drug delivery systems. These kinds of materials have important advantages since they are biodegradable, nontoxic, biologically compatible, and nonimmunogenic. Moreover, most polysaccharides have the advantage of having antimicrobial, anti-inflammatory, and mucoadhesion properties [11]. Therefore, HA, chitosan, dextran, and alginate are widely used in this kind of applications.

Since HA has been discovered in 1934 by scientists from Columbia University in New York, it has been clinically applied in numerous cases. HA is a polysaccharide composed of repetitive polymeric disaccharides of D-glucuronic acid and N-acetyl-D-glucosamine (GlcNAc) linked by β-bonds, which is ubiquitous in the human body and is crucial for many cellular and tissue functions (Figure 3.4). Its structure exhibits functional groups (−OH and −COOH) that promote and facilitate the conjugation of drugs via strong electrostatic interactions [8]. Moreover, these functional groups allow to modify HA flexibility and structure via chemical modifications of addition, condensation, or radical polymerization [1]. HA receptors are abundant in some specific tissues such as liver, kidney, and most tumor tissues, and HA is known to promote medicinal advantages when used, including tissue healing, cell proliferation and migration, angiogenesis, and inflammatory response control [15, 16].

Due to its physical, chemical, and biological characteristics, HA has been applied in drug delivery through ophthalmic, nasal, vaginal, pulmonary, parenteral, and topical routes [17]. By encapsulating recombinant human insulin with HA, a suitable dry powder for inhalation was obtained by spray drying, this allowed experiments for its inhalation into the lungs in beagle dogs. These insulin–HA microparticles, with 1–4 μm of diameter, prolonged the average retention time and the final half-life of the hormone [18].

In delivery through parental route, HA has been studied as a new protein and peptide drug carrier [19]. Each individual chain of HA can conjugate with a different protein or peptide, making multiple targets and drug effects possible [20]. Several products using HA as a carrier for tumor treatment are in development for potential future commercial products. Lee et al. synthesized amphiphilic HA-5β cholic acid polymer as an antitumor drug carrier, which effectively increased the uptake of tumor cells. Nanoparticles, with an average size between 250 and 350 nm, were accumulated more easily into tumor tissues than larger particles [21]. Galer et al. synthesized a HA–paclitaxel (PTX, a cytotoxic drug) conjugate (HA–PTX) in order to reduce the toxicity of PTX drugs family (derived from taxol) and improve the antitumor activity [22]. Recently, an Italian multinational company developed and patented a HA–PTX conjugate (Oncofid®P-B) for patients with bladder carcinoma, and it reached clinical trials. The

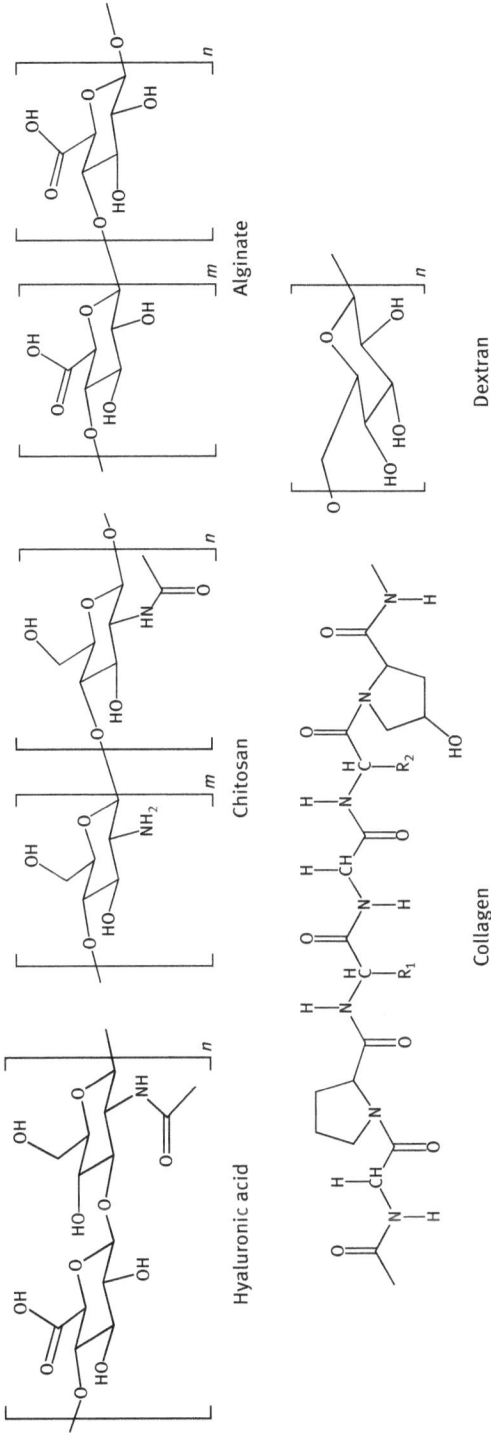

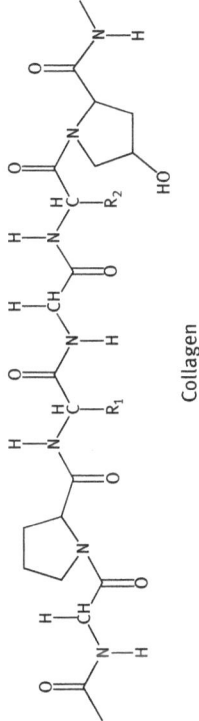

Figure 3.4: Chemical structure of the main natural polymers used in biomedical applications.

preliminary results to evaluate the safety, tolerability, and efficacy of the product were positive, showing minimal toxicity, absence of systemic absorption, excellent tolerability for prolonged treatment, and good efficacy [23].

Chitosan is a biopolymer composed of randomly distributed units of β-D-glucosamine and GlcNAc (Figure 3.4) and is produced by partial deacetylation of chitin. The degree of deacetylation modifies the interaction between chitosan and cells and the biodegradation of chitosan [24]. The primary hydroxyl and amine groups found on the backbone of chitosan allow chemical changes that modify its physical properties. As well as other natural polymers, chitosan is widely used due to its biocompatibility, biodegradability, nontoxicity, and resistance to chemical and biological degradation. Also, chitosan has a structure that exhibits functional groups that facilitate its modification and the adsorption of the biomolecules to deliver. Chitosan nano/microsized spherical particles are easy to obtain, and drugs or other bioactive substances such as proteins or enzymes can be incorporated inside them for their transport [24]. Furthermore, the mucosal adhesion and absorption characteristics of chitosan, which are due to their partial hydrophobic character, affect the rate of the drug release into the gastrointestinal tract, a fact that will in turn improve the bioavailability of various drugs [25]. Although chitosan was approved for dietary and wound dressing applications, it has not been approved for drug delivery in Europe and in the USA yet. However, several applications using chitosan nano/microparticles as carriers are in continuous development for potential future commercial products.

Several studies have shown the effective delivery of different anticancer drugs (cisplatin, 1-norcholesterol, PTX, and doxorubicin) when loaded into nano- and microparticles of chitosan [24, 26]. Moreover, chitosan nanoparticles were tested for delivery of protein/peptides and growth factors. For example, insulin showed better medical results and faster release when loaded into chitosan nanoparticles (size between 80 and 400 nm) compared with control solutions [27–29]. Likewise, carboxymethyl-β-cyclodextrin/chitosan nanoparticles were developed, and bovine serum albumin (BSA) was entrapped as a model protein for oral delivery. Results showed that this kind of nanoparticles (around 190 nm) based on chitosan constituted promising nanocarriers for oral delivery of protein drugs since the BSA release could be controlled in simulated gastric fluid, intestinal fluid, and colonic fluid [30]. In other studies, chondroitin sulfate immobilized onto chitosan hydrogel showed favorable results for cartilage formation [31], and the release of endothelial growth factor from chitosan–albumin microspheres (size between 400 and 600 μm) locally enhances the angiogenesis as evidenced by *in vivo* studies in rats [24]. Anti-inflammatory drugs were also loaded in chitosan particles (size between 1 and 2 mm), and their release was demonstrated so they could be potentially used for diseases such as arthritis, tendinitis, bursitis, and gout [32]. Moreover, antimicrobial peptides and nitric oxide were loaded in chitosan microparticles (size around 1.2 μm) and tested for their release, which demonstrated their potential use for the treatment of different infections from harmful bacteria [33].

Dextran is a polysaccharide formed by numerous glucose molecules. It has a linear 1,6-glycosidic bond with certain degree of branching through 1,3-linkage (Figure 3.4). Since dextran demonstrated biodegradability and biocompatibility in various organs of the human body, it is widely used in the field of medicine, particularly as a vehicle for drugs, proteins, bioactive agents, and as an antithrombotic agent [34]. In addition, it has numerous hydroxyl groups that support chemical modifications and alterations, allowing to conjugate different drugs and develop new types of dextran hydrogels, including glycidyl methacrylate dextran, hydroxyethyl methacrylate dextran, dextran hydroxyethyl methacrylate lactate, and dextran urethane [35]. Animal and human studies have shown that both the distribution and elimination of dextran are dependent on the molecular weight and electrostatic charge of this polymer. Drugs and proteins conjugated with dextran have resulted in increasing their half-life, preserving their therapeutic properties, altering their toxicity profile, and reducing the immunogenicity of drugs and/or proteins [36, 37]. Dextran nanoparticles and microspheres were conjugated with different anticancer or antiviral drugs obtaining promising results, especially for colon delivery, since their efficacies were higher compared to the free drugs [38–41].

Alginate is a natural polymer of (1→4′)-linked β-D-mannuronic acid and α-L-guluronic acid residues, which is biocompatible, easily available, and has high water sorption capacity [1, 42]. Since alginate does not induce a specific biological response, it has been used for biomedical applications such as wound dressings – with commercial products already in the market –, living cells encapsulation, and drug delivery [1, 3, 43]. The most common technique to form alginate particles is by ionic gelation using Ca^{2+} as counter ion of the negatively charged polymer chains. Usually, an alginate solution is dropping in a $CaCl_2$ solution to form particles. However, Madzovska-Malagurski et al. used Cu^{2+} instead of Ca^{2+} to develop Cu-alginate microbeads (size between 530 and 650 μm) in order to have a bactericidal effect against *Escherichia coli* and *Staphylococcus aureus* due to the presence of this cation. In addition, the authors demonstrated that the slow release of Cu^{2+} was suitable for promoting and maintaining chondrogenic phenotype of bovine calf chondrocytes in three-dimensional culture. These results demonstrated their potential for biomedical applications as part of antimicrobial wound dressing, tissue engineering scaffolds, or articular cartilage implants [42].

The use of alginate–$CaCl_2$ microspheres (size between 25 and 65 μm) was proposed as potential protein carriers for oral vaccine delivery [44]. Recently, alginate was used in combination with chitosan for oral mucosal immunization using hepatitis B antigen. Alginate-coated chitosan nanoparticles loaded with hepatitis B antigen (size around 605 nm) induce the immune response and the production of specific antibodies against this antigen (IgA at mucosal secretions and IgG antibodies in systemic circulation, in mice) [45]. Moreover, hepatitis A antigen using alginate–chitosan nanoparticles (size around 654 nm) as adjuvant/carrier improved the immunogenicity in mice by increasing the seroconversion rate, the hepatitis A antibodies level, and the splenocytes proliferation in comparison with conventional hepatitis A vaccine with alum [46]. In another approach, nanoparticles of alginate–dextran loaded with insulin were coated with chitosan–albumin and tested in oral delivery observing interaction with intestinal model cells. The developed system demonstrated clinical potential for the oral delivery of insulin and therapy of type 1 diabetes mellitus.

Alginate nanoparticles are widely studied for use in anticancer drugs delivery. In this sense, the development of alginate nanoparticles in combination with other polymers – especially chitosan – for cancer treatment has been described. Different cancer drugs such as bortezomib, curcumin, resveratrol, and doxorubicin were loaded in these nanoparticles and resulted in higher drug stability and less toxicity [47–51].

Hydrogel particles made of synthetic – hydrophobic or amphiphilic – polymers, such as PEG, polylactide-*co*-glycolide (PLGA), and polylactic acid (PLA) (Figure 3.5) are promising for drug delivery applications, although they have the disadvantage of requiring the use of organic solvents and surfactants for their synthesis, which can be harmful for cells. In addition, it can be necessary to remove the unreacted monomers or oligomers from the formed particles to avoid toxicity for clinical use [52]. The use of synthetic polymers is relevant in the cases where a hydrophobic drug with poor solubility has to be entrapped for drug delivery. PEG has negligible immunogenicity and toxicity and was approved by the U.S. Food and Drug Administration (FDA) for some pharmaceutical formulations as an excipient [8]. Peptides/proteins–PEG complexes have emerged as therapeutics treatment due to low immunogenicity and low clearance and pharmacokinetics [53]. Furthermore, peptide/protein loaded into polymeric nanoparticles has the advantage of protecting the therapeutic molecules from degradation. In this sense, Zhang et al. developed and tested insulin–PEG–chitosan nanoparticles (size between 150 and 300 nm) for insulin nasal administration in rabbits, and they observed better results than using insulin solution or suspension of PEG–chitosan [54].

In other examples, PLA and derivative nanoparticles were conjugated with various antitumor drugs such as docetaxel (DTX) and PTX [8]. However, for better results block copolymers can be formed. For example, PEG-b-PLGA nanoparticles (size between 60 and 140 nm) conjugated to therapeutic agent such as DTX were developed and were in phase II studies for the treatment of advanced or metastatic solid cancers [55, 56]. It is important to mention that PLGA and PLA were already approved for human use by the European Medicine Agency and FDA [57].

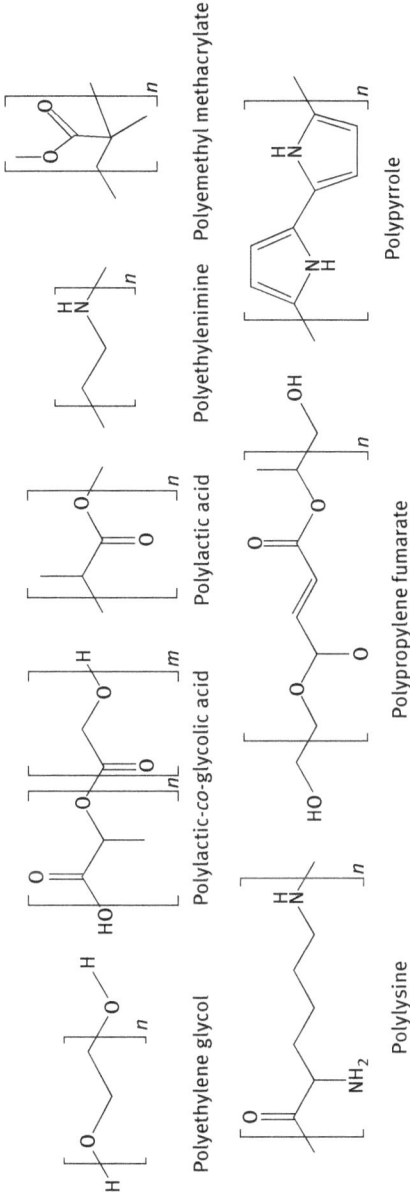

Figure 3.5: Chemical structure of the main synthetic polymers used in biomedical applications.

3.3.2 Gene carriers

The use of hydrogels as gene carriers improves the physiological stability of the DNA/RNA complex and safety; however, they have lower transfection efficiency than classical viral carriers. Some polymers such as polyethylenimine (PEI), polylysine (PLL), and chitosan have been already used for gene therapy for *in vitro* and *in vivo* experiments. In recent years, PLA and PLGA are other promising polymers as nonviral carriers for nucleic acid encapsulation [8].

PEI (Figure 3.5) is widely used and studied for gene transfection applications. PEI of high molecular weight have better transfection efficiency; however, its aggregation can cause cytotoxicity, and the DNA complex formation is affected by the polymer degree of branching [9, 58]. In order to overcome these drawbacks, PEI structure was chemically modified. The most utilized strategy is its combination with PEG to improve the *in vivo* half-life and cell viability [59]. A high density of short PEG chains conjugated to DNA, ribozymes, or oligonucleotides has better transfection rate, whereas a low density of longer PEG chains is more effective for double-stranded small interfering RNA (siRNA) [58]. Copolymer hydrogels of cross-linked PEI and PEO were reacted with polynucleotides [13]. This copolymer demonstrated good gene transfer activity. The modification of PEI nanoparticles with targeting ligands – like FA (a cellular penetrating peptide), TAT (the RGD peptide), or galactose – was also investigated to improve cell–particle interaction [59].

PLL (Figure 3.5) is a polypeptide formed with repeated units of amino acid lysine and tested *in vitro* and *in vivo* for delivery of genes [60]. PLLs with molecular weights greater than 3,000 Da are capable of forming a complex with DNA, but these complexes exhibit high cytotoxicity and form aggregates [58]. To overcome these disadvantages, derivatives or copolymers can be generated. For example, the addition of PEG increases the transfection efficiency and decreases the toxicity [8]. Lee et al. developed a PLL–PEG copolymer for the expression of an antisense mRNA using a plasmid and obtained excellent gene transfer results for both *in vitro* and *in vivo* experiments [61]. Likewise, Bikram et al. conjugated plasmid DNA with PLL–PEG nanoparticles (size between 150 and 200 nm) for gene delivery, obtaining better transfection efficiency [62]. In another study, Choi et al. synthesized and compared different PLL–PEG nanoparticles (size around 300 nm). The use of PLL–PEG nanoparticles demonstrated an increase in transfection efficiency using HepG2 cells compared to nanoparticles without PEG [63]. Park et al. also incorporated PLGA to PLL (size between 50 and 500 nm) to reduce cytotoxicity [64].

The effective use of chitosan for nonviral gene delivery has been demonstrated from various studies [52]. In this sense, it has been seen that the chitosan molecular weight, degree of deacetylation, chitosan/DNA ratio, and cell type affect the efficiency of transfection [24, 65]. Chitosan–DNA complexes are more stable using high-molecular-weight polymers, and an increased transduction efficiency was observed in correlation with a higher deacetylation degree [58]. Some authors studied the use of chitosan–plasmid DNA nanoparticles to treat hepatitis B. These nanoparticles (size between 300 and 400 nm) showed good efficacy *in vivo* after nasal mucosal administration [66]. Katas et al. developed and tested chitosan–siRNA nanoparticles (size between 280 and 710 nm) as a therapeutic agent to induce specific gene silencing mediated by siRNA. They explored chitosan as an siRNA vector due to its advantages such as low toxicity, biodegradability, and biocompatibility. *In vitro* studies using two different cells lines revealed that these nanoparticles succeeded in gene silencing and showed potential as viable vector for safer and cost-effective siRNA delivery [67]. Han et al. demonstrated the efficient use of chitosan nanoparticles (size around 200 nm) with Arg–Gly–Asp peptide (RGD peptide) attached for siRNA delivery system in animal models of ovarian cancer [68]. Likewise, several studies have used atelocollagen (type I collagen of calf dermis treated by pepsin) as a biomaterial for gene delivery (size between 100 and 300 nm) [69].

PLGA (Figure 3.5) nanoparticles are biocompatible, biodegradable, and widely used for *in vitro* and *in vivo* gene delivery [57]. DNA can be entrapped into or adsorbed onto the nanoparticles. For example, PLGA nanoparticles (600 nm size) with encapsulating plasmid DNA (with the reporter gene alkaline phosphatase) demonstrated high expression levels of protein product in *in vitro* and *in vivo* studies. Specifically, after intramuscular administration of these PLGA nanoparticles in rats, alkaline phosphatase protein expression (using as a model gene) increased at days 7 and 28. In other examples, PLGA nanoparticles (280 nm size) loaded with p53 gene DNA (a tumor suppressor protein), vascular endothelial growth factor DNA, or pigment epithelial-derived factor gene were tested in breast cancer cell line, for the treatment of ischemic heart disease or on mouse colon carcinoma cells *in vitro* and *in vivo*, respectively. These nanoparticles demonstrated a controlled release of the DNA, evidencing good results for therapeutic use [57]. Copolymerization of PLGA with other polymers such as chitosan and PEI has been reported to improve different characteristics. For example, PLGA–chitosan–plasmid DNA nanospheres (size around 60 nm) improved its cellular adsorption compared to nanospheres without chitosan [70]. In order to protect DNA against enzymatic degradation, PEI was incorporated into PLGA nanoparticles (size between 207 and 231 nm) and tested for gene expression in the human airway submucosal epithelial cell line Calu-3. Results demonstrated the presence of DNA inside the cells after 6 h of nanoparticle application [71].

3.3.3 Tissue engineering and scaffolds

Regenerative medicine provides the elements required for *in vivo* repair of damaged organs. In particular, tissue engineering seeks to repair and regenerate damaged tissue by combining a biodegradable matrix with living cells and/or biologically active molecules [72]. The natural or synthetic polymers used for tissue repair have different advantages and disadvantages. On the one hand, natural polymers (such as chitosan, alginate, and dextran) generally have low toxicity, good cell adhesion, and degradation and are economical but have low mechanical and chemical stability and can be rejected by the immune system. On the other hand, synthetic polymers (such as PLGA, poly ε-caprolactone (PCL), and polypropylene fumarate (PPF); Figure 3.5) have tunable physical and chemical properties and are easy to process but lack cell interaction properties. In addition, PLGA and PCL have the FDA approval for various medical applications [72]. For tissue repair, polymer nano/microparticles could be used in injectable formulations or alternatively can be impregnated as a part of the main polymer scaffold, which is not necessarily composed of the same material. The use of nano/microparticles tends to better control the release, improve the mechanical and porous properties, act as cell vehicles, and increase their biodegradability [9, 72].

Different growth factors were used in tissue engineering, such as bone morphogenetic proteins (BMP), transforming growth factor-beta (TGF-β), and insulin-like growth factor-I (IGF-I), for the differentiation of progenitor and/or stem cells [73]. Chen et al. developed microspheres of glycidylmethacrylated dextran–PEG with BMP loaded (size between 0.5 and 1.5 μm) for periodontal tissue regeneration. Microspheres that are used as part of scaffolds for periodontal therapy promoted the attachment, proliferation, and osteogenic differentiation of human periodontal ligament cells [74]. In addition, recombinant human BMP (rhBMP) combined with PLGA microspheres was tested *in vitro* demonstrating an enhancement of osteoblastic genes expression (in mouse pluripotent fibroblastic C3H10T1/2 cells and bone marrow stromal cells) and *in vivo* bone regeneration in rats and rabbit models [75–77]. Likewise, rhBMP was encapsulated into PLGA nanospheres (size around 300 nm) and then, these nanospheres were seeded onto a three-dimensional L-PLA scaffold, showing bone formation *in vivo* in rats and rhBMP delivery [78]. Niu et al. generated a scaffold of nanohydroxyapatite/collagen/L-PLA with chitosan microspheres (size between 10 and 60 μm) containing BMP-2 derived synthetic peptide. This novel microsphere–scaffold system demonstrated the stimulation of rabbit marrow mesenchymal stem cells during *in vitro* experiments [79].

In other possible applications, Jaklenec et al. developed scaffold containing PLGA microspheres with IGF-I and TGF-β1 encapsulated (size between 7 and 37 μm). The sequential release of these growth factors (for up to 70 days) make them useful for cartilage tissue engineering [73]. Furthermore, PLGA–PEG microparticles with recombinant human TGF-β (size between 19 and 23 μm) were tested for controlled release (up to 28 days) demonstrating to be a good vehicle for long-term delivery of this growth factor [80]. Similarly, Hedberg et al. synthesized and tested different PLGA–PEG microparticles with HA entrapped (17 μm size). The controlled release of HA from PLGA–PEG microparticles stimulates cell proliferation in various types of tissues for several days, demonstrating its potential use for tissue regeneration [81].

The use of smart hydrogels that respond to an electrical field has applications in neural prostheses and tissue engineering. This kind of polymers, like polypyrrole

(PPy), polyaniline, and polythiophene, promote cell growth and cell migration through electrical stimulation [82]. However, this class of polymers has the disadvantage of being nondegradable. In this way, several groups have developed degradable polymer materials, such as PLA and chitosan, with a low percentage of conducting polymers [82]. PPy (Figure 3.5) is a conducting polymer that has been shown to be biocompatible with cells and tissues *in vitro* and *in vivo*. Shi et al. synthesized membranes composed of the L-isomer of PLA and PPy nanoparticles (size between 100 and 300 nm) [83]. These membranes allowed cell adhesion and proliferation and also showed an increase in interleukins expression in the presence of electrical stimulation. Similarly, Huang et al. demonstrated Schwann cell adhesion and proliferation, as well as the expression and secretion of neurotrophic factors with electrical stimulation, using chitosan membranes with PPy nanoparticles (size around 30–120 nm) [84].

3.3.4 Biosensors using immobilized peptides and proteins

Bioresponsive or "smart" hydrogels are able to undergo structural modification (swelling or deswelling, degradation or erosion, and mechanical deformation) mediated by biological reactions. These structural modifications take place in response to different stimuli such as the increasing concentration of a specific biomolecule (protein, peptide, enzyme, antibody, etc), a change in the physiological environment or under pathological processes. [6]. In cellular environments, most stimuli-responsive mechanisms are under the fine control of different proteins, including enzymes. Thus, peptide/protein-responsive hydrogels can be a promising approach to respond directly to a target molecule and may provide important signals to monitor biological functions, detect physiological changes, or for the diagnosis of diseases.

Generally, three different types of strategies using responsive hydrogels could be found. The first type is related to the use of enzymes. Their immobilization into hydrogel particles could trigger a response according to the concentration of a substrate for the enzyme or changes in the surrounding environment. The second type of responsive hydrogels contain a short peptide, polysaccharide, or a molecule that can be programmed to respond to a specific enzyme. The third type is represented by antigen-responsive hydrogels. The immobilization of an antigen could be recognized by an antibody, and a specific response could be triggered [2].

An important example using enzymes for control release of insulin is the development of glucose-responsive hydrogels for diabetes treatment. Glucose oxidase enzyme (GOX) could be immobilized in different supports, as biosensors could compensate the inability of the pancreas to control glucose levels in the blood. The polymer used to entrap GOX in these cases showed pH responsiveness. When glucose concentration increases in the blood, GOX converts glucose to gluconic acid and H_2O_2, which lowers the pH inside the hydrogel. As a result, this pH reduction induces the ionization of functional groups and polymer chain repulsion inside the hydrogel network, leading to hydrogel swelling and insulin permeation throughout the network to reach the blood [85]. Sometimes, a catalase enzyme (CAT) forms part of the hydrogel particles to transform – and not accumulate – the H_2O_2 product of the GOX reaction in water and oxygen. Different materials have been used for insulin delivery by responding to a pH change. GOX immobilized on acryloyl cross-linked dextran dialdehyde nanoparticles (size between 48 and 74 nm) showed successful *in vitro* insulin release under artificial gastric fluid and artificial intestinal fluid conditions. These findings are promising to overcome problems related to subcutaneous insulin therapy [86]. In another example, PLGA nanoparticles (size below 200 nm) were prepared using a double-emulsion solvent diffusion method and were loaded with GOX and CAT enzymes. In addition, positively charged chitosan nanoparticles (size around 250 nm) were prepared using ionic gelation method and were loaded with insulin. Formulation of both nanoparticles and *in vivo* experiments using diabetic rats showed significant glycemic regulation up to 98 h after subcutaneous administration. These smart self-regulated drug delivery systems for insulin administration constitute desirable biosensors to achieve glycemic control and to decrease the long-term vascular complications in diabetes patients [87].

Another strategy to sense glucose levels includes the immobilization of lectins, such as concanavalin A (Con A) into hydrogel particles. Lectins are proteins that can recognize sugar moieties. This system exploits the higher affinity of Con A for glucose over the glycosylated insulin. The hydrogel contains Con A and glycosylated insulin, which binds to the lectin via its sugar moiety. When free glucose is detected, glycosylated insulin would be released from inside of the hydrogel by diffusion [85, 88].

The last mechanism of glucose-responsive hydrogels is based on phenylboronic acid (PBA). PBA has affinity for polyol molecules (sugar alcohols) and, therefore, could sense free glucose. Phenylborate groups exhibit an equilibrium between uncharged and charged forms. Since the reaction between PBA and glucose occurs through the cationic form of the PBA, the equilibrium is displaced toward the cationic charged species. The increasing charge density in the hydrogel promotes polymer chains repulsion and increases the hydrophilicity, leading to swelling and therefore the release of the insulin from inside the hydrogel. PBA could be linked to a determined polymer to form hydrogel particles. For example, porous PLGA microspheres (size around 10 μm) were prepared and loaded with insulin. Then microspheres were coated using PVA and a novel boronic acid-containing copolymer – poly(acrylamide phenyl boronic acid-*co*-*N*-vinylcaprolactam)–. The insulin-loaded microspheres could regulate drug release in response to varying glucose concentration in *in vitro* experiments. In *in vivo* studies using diabetic mice model, these smart microspheres showed effective control of blood sugar level over at least 18 days, retaining their glucose-sensitive properties during this time [89].

Some responsive hydrogels that contain a short peptide, polysaccharide, or other molecule are also widely exploited to release some drugs in response to a specific enzyme. For example, some PEG-based hydrogels are cross-linked using a metalloproteinase-sensitive peptide (MMP). The metalloproteinases are a family of proteases that are normally upregulated in cartilage repair, promoting the chondrocyte differentiation [90, 91]. In addition, metalloproteinases are present in tumor tissues, and this kind of strategy could be used to deliver a specific drug inside the MMP–PEG particles. MMP–PEG particles will be degraded by metalloproteinases, and encapsulated drugs could be liberated specifically in tumor tissues [92, 93].

Other responsive hydrogels are exploited to release drugs specifically in the colon. The use of pH-sensitive monomers (acrylamide derivatives and acrylic acid) cross-linked with azo-aromatic bonds allows the degradability of the system to be restricted to the colon environment. When the particles arrive to the colon, the hydrogels reach their highest swelling degree. Then, the cross-link bonds are degraded by the action of azo-reductase enzymes or mediators, and the drug can be released. Other biodegradable polymers that are enzyme cleavable are promising materials for developing site-specific drug release systems by enzymatic decomposition. Naturally derived polymers such as pectin, amylose, gellan gum, chitosan, chondroitin sulfate, alginate, and dextran or synthetic polymers such as PEG, PEO, pHEMA, and poly(N-isopropylacrylamide) are in continuous study for this kind of biomedical applications [2].

The third strategy to develop responsive hydrogels involves the use of antibodies and biomolecules that can recognize specific antigens. In the presence of an antigen, hydrogels with immobilized antibodies in their structure are able to undergo structural or volume changes as a consequence of the antigen–antibody interaction. Several polymers were used for coupling different antibodies and to develop this kind of biosensors. The system swelling was dependent on the antigen concentration, pH, and temperature [94–97].

In the last decade, in addition to the use of biosensors for disease treatment as discussed earlier, the development of nanobiosensors has seen great advancements for the characterization and quantification of biomolecules to improve clinical diagnosis of genetic and infectious diseases. Different types of electrochemical biosensors based their detection on amperometric changes due to a specific enzyme reaction or protein–ligand interaction. The immobilization of proteins directly in a metallic electrode often promotes their denaturalization and leads to the impossibility of reusing them, while the deposition of proteins previously immobilized onto hydrogel particles – especially using chitosan – avoids their denaturation, allows their reuse, and increases the sensitivity due to the high surface-area-to-volume ratio of the hydrogel nanoparticles [98].

Finally, it is important to remark that all the peptides, proteins, antibodies, and enzymes used for medical treatment and diagnosis discussed earlier are initially purified using hydrogel particles. Commercial chromatographic supports are based on cross-linked and modified agarose, dextran, cellulose, or PAAm hydrogel particles. For their commercialization, these materials have different particle sizes, ranging from 15 to 300 μm, and are usually porous, to give a high internal surface area. This area of research is in continuous development and the use of new hydrogels – such as chitosan, alginate, and methacrylate derivatives – in the form of nano- and microparticles and in other forms is proposed for the purification of numerous biomolecules with biomedical applications [99–104].

3.3.5 Cosmetics

Since the introduction of the first filler material approved by the FDA – the bovine collagen – 30 years ago, the studies in this field have increased, and nowadays, there are numerous filler producer companies around the world. Global dermal filler market will reach USD 8.5 billion by 2024, probably due to an increase in the demand of antiaging and wrinkle treatments [3].

Biodegradable fillers have been used to adjust facial soft tissue defects. They are eventually metabolized by the body, usually in a period ranging from months to a year, offering safety without health complications. Biodegradable fillers can be divided into two main classes, according to their duration in the human body: the nonpermanent fillers, such as collagen or HA, and the semipermanent fillers, such as PLA and calcium hydroxyapatite [105]. The most used filler hydrogels in cosmetic industry are made of bovine collagen and HA (Figure 3.4), both biodegradable. PAAm hydrogel, a nonbiodegradable filler, is also used when a long-lasting effect is desired. The hydrogel concentration in the filler determines the longevity and the stability of the correction intervention. Gels with higher stiffness can provide a better support in facial muscles and they better resist the dynamic forces acting during their movements. On the other side, gels with low modulus are more suitable for areas with static and superficial wrinkles, where the mechanical resistance is not a critical factor. Rheological behavior and particles size affect the flow of the filler passing through a syringe. Hydrogels tend to swell post injection; thus, this behavior has to be considered before the application [106].

Usually, the filler formulation provided by the manufacturers includes both cross-linked gels and a fluid component of the hydrogel. This latter one is easily metabolized by the human body and does not contribute to the duration and effectiveness of the product. Furthermore, some commercial products include hydrogels in their formulation in the form of nano- and microparticles that help to maintain the desired results much longer and also stimulate natural collagen and HA production [58, 107].

Collagen is the major structural component of the dermis, and it has a fundamental role in providing strength and support to human skin [108]. For example, the product Artefll® (known in Europe as Artecoll®) is a mixture of bovine collagen and homogeneous PMMA (Figure 3.5) microspheres. This product helps to maintain the desired results much longer and also stimulate natural collagen production. In the case of HA, it is commonly adopted by elderly people to correct facial lines and reduce wrinkling. Restylane® and Perlane® products are nonanimal-stabilized HAs obtained from bacterial cultures. Both of them are cross-linked with butanediol diglycidyl ether. Restylane® was the first nonanimal-stabilized HA approved in the USA in 2003 [106]; Perlane® was approved later, in 2007. The only difference between these two products is the particle size: the largest fraction of gel particles for Perlane® is between 940 and 1,090 µm, whereas the largest fraction for Restylane® is between 250 and 500 µm. HA fillers have the advantage over collagen of being instantly reversible by the application of hyaluronidase, an enzyme that degrades HA [3, 109].

3.3.6 Diagnostic imaging

Another interesting biomedical application developed in the recent years that involves the use of hydrogel particles is diagnostic imaging. Polymeric particles could be used as bioimaging probes for diagnostic applications improving the imaging signals, protecting the contrast agent, and ensuring the correct body or tissue distribution [8]. Polymeric bioimaging probes have prolonged half-life, enhanced stability, reduced toxicity, and improved targeting, along with reduced nonspecific binding [110]. Polymer-conjugated probes can prolong blood retention time resulting in better images. However, it is necessary to use biodegradable polymers to avoid toxicity for the possible incomplete probe clearance [111].

Magnetic resonance imaging (MRI) is used in medical settings to produce high-quality images of the internal organs. This technique measures the changes in the characteristics of hydrogen nuclei in water and other nuclei with similar chemical shift across the image slice. Bioimaging probes – in combination or not with other contrast agents – could be used in MRI to enhance the contrast between soft tissues. Typically contrast agents like gadolinium (Gd) and iron oxide were conjugated with different biocompatible polymers, such as chitosan, dextran, polyamine, PEG, and/or PLL [26, 111]. Gd conjugated with PLL particles with various molecular weights are already commercially available [111]. Chitosan microspheres loaded with Gd diethylenetriaminopenta acetic acid (size around 11.7 µm) were tested and showed good results of *in vitro* Gd release that are promising for its application in MRI [112]. Likewise, iron oxide conjugated to dextran or carboxydextran nanoparticles are commercially available as nanoparticles with the name of Feridex® (size between 50 and 200 nm) and Resovist® (size around 60 nm), respectively [111]. In all the cases exemplified earlier, the combination of the metal ion with these polymers as a coating can decrease their toxicity, prolong their presence in circulation, and improve tissue specificity.

For nuclear imaging, such as single-photoemission computed tomography and positron emission tomography, radioactive agents are commonly used and their combination with polymers – such as PLL, dextran, and PEG – improved the specificity, prolonged the circulating time, and amplified the signal [111].

The most commonly used X-ray computed tomography contrast agent is the low-molecular-weight iodine. The probe can be attached to polymeric nanoparticles of PLA–PEG and/or PLGA–PEG in order to reduce the acquisition image time and the heavy atom toxicity [9, 113].

Indocyanine green and quantum dots are used as contrasts agents for fluorescence imaging [111]. Polymers conjugated to fluorescent agents solve the problems of short circulation time and lack of specificity [8]. One of the polymers used for this purpose is PEG–PLL copolymer [111].

3.4 Future remarks

Hydrogels have been used in biomedical applications since the 1960s, and during these years, it attracted the attention of numerous researchers. The studies devoted to the development of new hydrogel applications are exponentially increasing, and uncountable hydrogel modification using synthetic, natural, or blend polymers are in constant development. However, only a small part of these developments has resulted in the invention of a new product for people's daily life. Many processing techniques concerning hydrogels that are performed at lab scale are difficult, or even impossible to be reproduced/implemented at industrial scale. Moreover, a deep polymer characterization concerning nature, molecular weight, and chemical modification, among others is necessary to validate homogeneity of the initial material in each batch of production. These requirements are sometimes difficult to implement in hydrogels development using natural polymers, which generally present high polydispersity. Finally, the type of chemical cross-linker used usually implies a rigorous process of validation of its absence in the final product. Sometimes these and other requirements are not considered during the academic development of new products, and this fact would impair their approval for commercial purposes by the pertinent authorities.

References

[1]   Gyles D.A., Castro L.D., Silva J.O.C., Ribeiro-Costa R.M. A review of the designs and prominent biomedical advances of natural and synthetic hydrogel formulations. Eur Polym J 2017, 88, 373–392.

[2]   Ferreira N.N., Ferreira L.M.B., Cardoso V.M.O., Boni F.I., Souza A.L.R., Gremião M.P.D. Recent advances in smart hydrogels for biomedical applications: From self-assembly to functional approaches. Eur Polym J 2018, 99, 117–133.

[3]   Cascone S., Lamberti G. Hydrogel-based commercial products for biomedical applications: A review. Int J Pharm 2020, 573, 118803.

[4]   Hoffman A.S. Hydrogels for biomedical applications. Adv Drug Deliv Rev 2012, 64, 18–23.

[5]   Batista R.A., Espitia P.J.P., Quintans J.D.S.S. et al. Hydrogel as an alternative structure for food packaging systems. Carbohydr Polym 2019, 205, 106–116.

[6]   Zhao W., Jin X., Cong Y., Liu Y., Fu J. Degradable natural polymer hydrogels for articular cartilage tissue engineering. J Chem Technol Biot 2013, 88(3), 327–339.

[7]   Amsden B. Solute Diffusion within Hydrogels. Mechanisms and Models. Macromolecules 1998, 31(23), 8382–8395.

[8]   Tang Z., He C., Tian H. et al. Polymeric nanostructured materials for biomedical applications. Prog Polym Sci 2016, 60, 86–128.

[9]   Vauthier C., Bouchemal K. Methods for the preparation and manufacture of polymeric nanoparticles. Pharm Res 2009, 26(5), 1025–1058.

[10]   Lin -C.-C., Metters A.T. Hydrogels in controlled release formulations: network design and mathematical modeling. Adv Drug Deliv Rev 2006, 58(12), 1379–1408.

[11]   Dragan E.S., Dinu M.V. Polysaccharides constructed hydrogels as vehicles for proteins and peptides. A review Carbohyd Polym 2019, 225, 115210.

[12]   Barua S., Mitragotri S. Challenges associated with penetration of nanoparticles across cell and tissue barriers: A Review of current status and future. Nano today 2014, 9(2), 223–243. ProspectsDragan ES, Dinu MV. Polysaccharides constructed hydrogels as vehicles for proteins and peptides. A review. Carbohyd Polym 2019, 225, 115210.

[13]   Vinogradov S.V., Bronich T.K., Kabanov A.V. Nanosized cationic hydrogels for drug delivery: preparation, properties and interactions with cells. Adv Drug Deliv Rev 2002, 54(1), 135–147.

[14]   Barclay T.G., Day C.M., Petrovsky N., Garg S. Review of polysaccharide particle-based functional drug delivery. Carbohyd Polym 2019, 221, 94–112.

[15]   Burdick J.A., Prestwich G.D. Hyaluronic acid hydrogels for biomedical applications. Adv Mater 2011, 23(12), H41–H56.

[16]   Rosso F., Quagliariello V., Tortora C., Di Lazzaro A., Barbarisi A., Laffaioli R.V. Cross-linked hyaluronic acid sub-micron particles: in vitro and in vivo biodistribution study in cancer xenograft model. J Mater Sci-Mater M 2013, 24(6), 1473–1481.

[17]   Huang G., Huang H. Application of hyaluronic acid as carriers in drug delivery. Drug Deliv 2018, 25(1), 766–772.

[18]   Surendrakumar K., Martyn G.P., Hodgers E.C.M., Jansen M., Blair J.A. Sustained release of insulin from sodium hyaluronate based dry powder formulations after pulmonary delivery to beagle dogs. J Control Release 2003, 91(3), 385–394.

[19]   Goodarzi N., Varshochian R., Kamalinia G., Atyabi F., Dinarvand R. A review of polysaccharide cytotoxic drug conjugates for cancer therapy. Carbohyd Polym 2013, 92(2), 1280–1293.

[20]   Jiang T., Zhang Z., Zhang Y. et al. Dual-functional liposomes based on pH-responsive cell-penetrating peptide and hyaluronic acid for tumor-targeted anticancer drug delivery, Biomaterials 2012, 33(36), 9246–9258.

[21]   Lee D.-E., Kim A.Y., Yoon H.Y. et al. Amphiphilic hyaluronic acid-based nanoparticles for tumor-specific optical/MR dual imaging, J Mater Chem 2012, 22(21), 10444–10447.

[22]   Galer C.E., Sano D., Ghosh S.C. et al. Hyaluronic acid–paclitaxel conjugate inhibits growth of human squamous cell carcinomas of the head and neck via a hyaluronic acid-mediated mechanism, Oral Oncol 2011, 47(11), 1039–1047.

[23] Bassi P.F., Volpe A., D'agostino D. et al. Paclitaxel-hyaluronic acid for intravesical therapy of bacillus calmette-guérin refractory carcinoma in situ of the bladder: results of a phase I study, J Urol 2011, 185(2), 445–449.

[24] Dash M., Chiellini F., Ottenbrite R.M., Chiellini E. Chitosan – A versatile semi-synthetic polymer in biomedical applications. Prog Polym Sci 2011, 36(8), 981–1014.

[25] Rani M., Agarwal A., Negi Y.S. Review: chitosan based hydrogel polymeric beads – as drug delivery system. BioResources 2010, 5, 2765–2767.

[26] Park J.H., Saravanakumar G., Kim K., Kwon I.C. Targeted delivery of low molecular drugs using chitosan and its derivatives. Adv Drug Deliv Rev 2010, 62(1), 28–41.

[27] Grenha A., Seijo B., Remuñán-López C. Microencapsulated chitosan nanoparticles for lung protein delivery. Eur J Pharm 2005, 25(4), 427–437.

[28] Fernández-Urrusuno R., Calvo P., Remuñán-López C., Vila-Jato J.L., José Alonso M. Enhancement of nasal absorption of insulin using chitosan nanoparticles. Pharm Res 1999, 16(10), 1576–1581.

[29] Erel G., Kotmakçı M., Akbaba H., Sözer Karadağlı S., Kantarcı A.G. Nanoencapsulated chitosan nanoparticles in emulsion-based oral delivery system: In vitro and in vivo evaluation of insulin loaded formulation. J Drug Deliv Sci Tec 2016, 36, 161–167.

[30] Song M., Li L., Zhang Y., Chen K., Wang H., Gong R. Carboxymethyl-β-cyclodextrin grafted chitosan nanoparticles as oral delivery carrier of protein drugs. React Funct Polym 2017, 117, 10–15.

[31] Francis Suh J.K., Matthew H.W.T. Application of chitosan-based polysaccharide biomaterials in cartilage tissue engineering: a review. Biomaterials 2000, 21(24), 2589–2598.

[32] Shiraishi S., Imai T., Otagiri M. Controlled release of indomethacin by chitosan-polyelectrolyte complex: optimization and in vivo/in vitro evaluation. J Control Release 1993, 25(3), 217–225.

[33] Sun Y., Liu Y., Liu W., Lu C., Wang L. Chitosan microparticles ionically cross-linked with poly(γ-glutamic acid) as antimicrobial peptides and nitric oxide delivery systems. Biochem Eng J 2015, 95, 78–85.

[34] Li H., Niu R., Yang J., Nie J., Yang D. Photocrosslinkable tissue adhesive based on dextran. Carbohyd Polym 2011, 86(4), 1578–1585.

[35] Stenekes R.J.H., Talsma H., Hennink W.E. Formation of dextran hydrogels by crystallization. Biomaterials 2001, 22(13), 1891–1898.

[36] Mehvar R. Dextrans for targeted and sustained delivery of therapeutic and imaging agents. J Control Release 2000, 69(1), 1–25.

[37] Bae K.H., Lee F., Xu K. et al. Microstructured dextran hydrogels for burst-free sustained release of PEGylated protein drugs. Biomaterials 2015, 63, 146–157.

[38] Sugahara S.-I., Kajiki M., Kuriyama H., Kobayashi T.-R. Complete regression of xenografted human carcinomas by a paclitaxel–carboxymethyl dextran conjugate (AZ10992). J Control Release 2007, 117(1), 40–50.

[39] Tiryaki E., Başaran Elalmış Y., Karakuzu İkizler B., Yücel S. Novel organic/inorganic hybrid nanoparticles as enzyme-triggered drug delivery systems: Dextran and Dextran aldehyde coated silica aerogels. J Drug Deliv Sci Tec 2020, 56, 101517.

[40] Joshy K.S., George A., Snigdha S. et al. Novel core-shell dextran hybrid nanosystem for anti-viral drug delivery. Mater Sci Eng C 2018, 93, 864–872.

[41] Chen F., Huang G., Huang H. Preparation and application of dextran and its derivatives as carriers. Int J Biol Macromol 2020, 145, 827–834.

[42] Madzovska-Malagurski I., Vukasinovic-Sekulic M., Kostic D., Levic S. Towards antimicrobial yet bioactive Cu-alginate hydrogels. Biomed Mater 2016, 11(3), 035015.

[43] Queen D., Orsted H., Sanada H., Sussman G. A dressing history. Int Wound J 2004, 1(1), 59–77.
[44] Hariyadi D.M., Ma Y., Wang Y. et al. The potential for production of freeze-dried oral vaccines using alginate hydrogel microspheres as protein carriers, J Drug Deliv Sci Tec 2014, 24(2), 178–184.
[45] Saraf S., Jain S., Sahoo R.N., Mallick S. Lipopolysaccharide derived alginate coated Hepatitis B antigen loaded chitosan nanoparticles for oral mucosal immunization. Int J Biol Macromol 2020, 154, 466–476.
[46] Abdelallah N.H., Gaber Y., Rashed M.E., Azmy A.F., Abou-Taleb H.A., Abdelghani S. Alginate-coated chitosan nanoparticles act as effective adjuvant for hepatitis A vaccine in mice. Int J Biol Macromol 2020, 152, 904–912.
[47] Dheer D., Arora D., Jaglan S., Rawal R.K., Shankar R. Polysaccharides based nanomaterials for targeted anti-cancer drug delivery. J Drug Target 2017, 25(1), 1–16.
[48] Das R.K., Kasoju N., Bora U. Encapsulation of curcumin in alginate-chitosan-pluronic composite nanoparticles for delivery to cancer cells. Nanomed-Nanotechnol 2010, 6(1), 153–160.
[49] Rezk A.I., Obiweluozor F.O., Choukrani G., Park C.H., Kim C.S. Drug release and kinetic models of anticancer drug (BTZ) from a pH-responsive alginate polydopamine hydrogel: Towards cancer chemotherapy. Int J Biol Macromol 2019, 141, 388–400.
[50] Nazlı A.B., Açıkel Y.S. Loading of cancer drug resveratrol to pH-sensitive, smart, alginate-chitosan hydrogels and investigation of controlled release kinetics. J Drug Deliv Sci Tec 2019, 53, 101199.
[51] Elbialy N.S., Mohamed N. Alginate-coated caseinate nanoparticles for doxorubicin delivery: Preparation, characterisation, and in vivo assessment. Int J Biol Macromol 2020, 154, 114–122.
[52] Prabaharan M., Mano J.F. Chitosan-based particles as controlled drug delivery systems. Drug Deliv 2004, 12(1), 41–57.
[53] Kontos S., Hubbell J.A. Drug development: longer-lived proteins. Chem Soc Rev 2012, 41(7), 2686–2695.
[54] Zhang X., Zhang H., Wu Z., Wang Z., Niu H., Li C. Nasal absorption enhancement of insulin using PEG-grafted chitosan nanoparticles. Eur J Pharm Biopharm 2008, 68(3), 526–534.
[55] Autio K.A., Dreicer R., Anderson J. el al Safety and efficacy of BIND-014, a docetaxel nanoparticle targeting prostate-specific membrane antigen for patients with metastatic castration-resistant prostate cancer: a phase 2 clinical trial, JAMA Oncol 2018, 4(10), 1344–1351.
[56] Zale S.E., Troiano G., Ali M.M., Hrkach J., Wright J. Drug Loaded Polymeric Nanoparticles and Methods of Making and Using Same. (Cambridge, MA, US): United States, BIND Biosciences, Inc., 2012.
[57] Nimesh S. 15 – Poly(D,L-lactide-co-glycolide)-based nanoparticles. In Nimesh S., ed. Gene Therapy. Sawston, Cambridge, UK, Woodhead Publishing, 2013, 309–329.
[58] Mintzer M.A., Simanek E.E. Nonviral vectors for gene delivery. Chem Rev 2009, 109(2), 259–302.
[59] Nimesh S. 10 – Polyethylenimine nanoparticles. In Nimesh S., ed. Gene Therapy. Sawston, Cambridge, UK, Woodhead Publishing, 2013, 197–223.
[60] Nimesh S. 8 – Poly-L-lysine nanoparticles. In Nimesh S., ed. Gene Therapy. Sawston, Cambridge, UK, Woodhead Publishing, 2013, 147–162.
[61] Lee M., Han S.-O., Ko K.S. el al Repression of GAD autoantigen expression in pancreas β-cells by delivery of antisense plasmid/PEG-g-PLL complex, Mol Ther 2001, 4(4), 339–346.

[62] Bikram M., Ahn C.-H., Chae S.Y., Lee M., Yockman J.W., Kim S.W. Biodegradable poly (ethylene glycol)-co-poly(l-lysine)-g-histidine multiblock copolymers for nonviral gene delivery. Macromolecules 2004, 37(5), 1903–1916.

[63] Choi Y.H., Liu F., Kim J.-S., Choi Y.K., Jong Sang P., Kim S.W. Polyethylene glycol-grafted poly-l-lysine as polymeric gene carrier. J Control Release 1998, 54(1), 39–48.

[64] Park S., Healy K.E. Nanoparticulate DNA packaging using terpolymers of poly(lysine-g-(lactide-b-ethylene glycol)). Bioconjug Chem 2003, 14(2), 311–319.

[65] Nimesh S. 9 – Chitosan nanoparticles. In Nimesh S., ed. Gene Therapy. Sawston, Cambridge, UK, Woodhead Publishing, 2013, 163–196.

[66] Khatri K., Goyal A.K., Gupta P.N., Mishra N., Vyas S.P. Plasmid DNA loaded chitosan nanoparticles for nasal mucosal immunization against hepatitis B. Int J Pharm 2008, 354(1), 235–241.

[67] Katas H., Alpar H.O. Development and characterisation of chitosan nanoparticles for siRNA delivery. J Control Release 2006, 115(2), 216–225.

[68] Han H.D., Mangala L.S., Lee J.W. et al. Targeted gene silencing using RGD-labeled chitosan nanoparticles, Clin Cancer Res 2010, 16(15), 3910.

[69] Nimesh S. 11 – Atelocollagen. In Nimesh S., ed. Gene Therapy. Sawston, Cambridge, UK, Woodhead Publishing, 2013, 225–235.

[70] Zeng P., Xu Y., Zeng C., Ren H., Peng M. Chitosan-modified poly(d,l-lactide-co-glycolide) nanospheres for plasmid DNA delivery and HBV gene-silencing. Int J Pharm 2011, 415,(1), 259–266.

[71] Bivas-Benita M., Romeijn S., Junginger H.E., Borchard G. PLGA–PEI nanoparticles for gene delivery to pulmonary epithelium. Eur J Pharm Biopharm 2004, 58(1), 1–6.

[72] Puppi D., Chiellini F., Piras A.M., Chiellini E. Polymeric materials for bone and cartilage repair. Prog Polym Sci 2010, 35(4), 403–440.

[73] Jaklenec A., Hinckfuss A., Bilgen B., Ciombor D.M., Aaron R., Mathiowitz E. Sequential release of bioactive IGF-I and TGF-β1 from PLGA microsphere-based scaffolds. Biomaterials 2008, 29(10), 1518–1525.

[74] Chen F.-M., Zhao Y.-M., Zhang R. et al. Periodontal regeneration using novel glycidyl methacrylated dextran (Dex-GMA)/gelatin scaffolds containing microspheres loaded with bone morphogenetic proteins, J Control Release 2007, 121(1), 81–90.

[75] Puleo D.A., Huh W.W., Duggirala S.S., Deluca P.P. In vitro cellular responses to bioerodible particles loaded with recombinant human bone morphogenetic protein-2. J Biomed Mater Res 1998, 41(1), 104–110.

[76] Smith J.L., Jin L., Parsons T. et al. Osseous regeneration in preclinical models using bioabsorbable delivery technology for recombinant human bone morphogenetic protein 2 (rhBMP-2), J Control Release 1995, 36(1), 183–195.

[77] Woo B.H., Fink B.F., Page R. et al. Enhancement of bone growth by sustained delivery of recombinant human bone morphogenetic protein-2 in a polymeric matrix, Pharm Res 2001, 18(12), 1747–1753.

[78] Wei G., Jin Q., Giannobile W.V., Ma P.X. The enhancement of osteogenesis by nano-fibrous scaffolds incorporating rhBMP-7 nanospheres. Biomaterials 2007, 28(12), 2087–2096.

[79] Niu X., Feng Q., Wang M., Guo X., Zheng Q. Porous nano-HA/collagen/PLLA scaffold containing chitosan microspheres for controlled delivery of synthetic peptide derived from BMP-2. J Control Release 2009, 134(2), 111–117.

[80] Lu L., Stamatas G.N., Mikos A.G. Controlled release of transforming growth factor β1 from biodegradable polymer microparticles. J Biomed Mater Res 2000, 50(3), 440–451.

[81] Hedberg E.L., Shih C.K., Solchaga L.A., Caplan A.I., Mikos A.G. Controlled release of hyaluronan oligomers from biodegradable polymeric microparticle carriers. J Control Release 2004, 100(2), 257–266.

[82] Guo B., Glavas L., Albertsson A.-C. Biodegradable and electrically conducting polymers for biomedical applications. Prog Polym Sci 2013, 38,(9), 1263–1286.

[83] Shi G., Zhang Z., Rouabhia M. The regulation of cell functions electrically using biodegradable polypyrrole–polylactide conductors. Biomaterials 2008, 29(28), 3792–3798.

[84] Huang J., Hu X., Lu L., Ye Z., Zhang Q., Luo Z. Electrical regulation of Schwann cells using conductive polypyrrole/chitosan polymers. J Biomed Mater Res A 2010, 93A(1), 164–174.

[85] Shen D., Yu H., Wang L. et al. Recent progress in design and preparation of glucose-responsive insulin delivery systems. J Control Release 2020, 321, 236–258.

[86] Jamwal S., Ram B., Ranote S., Dharela R., Chauhan G.S. New glucose oxidase-immobilized stimuli-responsive dextran nanoparticles for insulin delivery. Int J Biol Macromol 2019, 123, 968–978.

[87] Mohammadpour F., Hadizadeh F., Tafaghodi M. et al. Preparation, in vitro and in vivo evaluation of PLGA/Chitosan based nano-complex as a novel insulin delivery formulation. Int J Pharm 2019, 572, 118710.

[88] Yin R., Bai M., He J., Nie J., Zhang W. Concanavalin A-sugar affinity based system: Binding interactions, principle of glucose-responsiveness, and modulated insulin release for diabetes care. Int J Biol Macromol 2019, 124, 724–732.

[89] Wu J.-Z., Williams G.R., Li H.-Y., Wang D.-X., Li S.-D., Zhu L.-M. Insulin-loaded PLGA microspheres for glucose-responsive release. Drug Deliv 2017, 24(1), 1513–1525.

[90] Park Y., Lutolf M.P., Hubbell J.A., Hunziker E.B., Wong M. Bovine primary chondrocyte culture in synthetic matrix metalloproteinase-sensitive poly(ethylene glycol)-based hydrogels as a scaffold for cartilage repair. Tissue Eng 2004, 10(3–4), 515–522.

[91] Ren Y., Zhang H., Qin W., Du B., Liu L., Yang J. A collagen mimetic peptide-modified hyaluronic acid hydrogel system with enzymatically mediated degradation for mesenchymal stem cell differentiation. Mater Sci Eng C 2020, 108, 110276.

[92] Ruan S., Cao X., Cun X. et al. Matrix metalloproteinase-sensitive size-shrinkable nanoparticles for deep tumor penetration and pH triggered doxorubicin release. Biomaterials 2015, 60, 100–110.

[93] Xu J.-H., Gao F.-P., Li -L.-L. et al. Gelatin–mesoporous silica nanoparticles as matrix metalloproteinases-degradable drug delivery systems in vivo. Microporous Mesoporous Mater 2013, 182, 165–172.

[94] Lim S.L., Ooi C.-W., Tan W.S., Chan E.-S., Ho K.L., Tey B.T. Biosensing of hepatitis B antigen with poly(acrylic acid) hydrogel immobilized with antigens and antibodies. Sens Actuators B Chem 2017, 252, 409–417.

[95] Lim S.L., Ooi C.-W., Low L.E., Tan W.S., Chan E.-S., Ho K.L., Tey B.T. Synthesis of poly (acrylamide)-based hydrogel for bio-sensing of hepatitis B core antigen. Mater Chem Phys 2020, 243, 122578.

[96] Miyata T., Asami N., Uragami T. A reversibly antigen-responsive hydrogel. Nature 1999, 399 (6738), 766–769.

[97] Lu Z.-R., Kopečková P., Kopeček J. Antigen responsive hydrogels based on polymerizable antibody fab' fragment. Macromol Biosci 2003, 3, 296–300.

[98] Hasanzadeh M., Shadjou N., De La Guardia M. Electrochemical biosensing using hydrogel nanoparticles. Trac-Trends Anal Chem 2014, 62, 11–19.

[99] Li Y., Stern D., Lock L.L. et al. Emerging biomaterials for downstream manufacturing of therapeutic proteins. Acta Biomater 2019, 95, 73–90.

[100] Baieli M.F., Urtasun N., Miranda M.V., Cascone O., Wolman F.J. Isolation of lactoferrin from whey by dye-affinity chromatography with Yellow HE-4R attached to chitosan mini-spheres. Int Dairy J 2014, 39(1), 53–59.

[101] Urtasun N., Baieli M.F., Cascone O., Wolman F.J., Miranda M.V. High-level expression and purification of recombinant wheat germ agglutinin in Rachiplusia nu larvae. Process Biochem 2015, 50(1), 40–47.

[102] Jain S., Mondal K., Gupta M.N. Applications of alginate in bioseparation of proteins. Artif Cells Blood Sub 2006, 34(2), 127–144.

[103] Shao M., Xiu L., Zhang H., Huang J., Gong X. Chitosan/cellulose-based beads for the affinity purification of histidine-tagged proteins. Prep Biochem Biotech 2018, 48(4), 352–360.

[104] Baieli M.F., Urtasun N., Miranda M.V., Cascone O., Wolman F.J. Efficient wheat germ agglutinin purification with a chitosan-based affinity chromatographic matrix. J Sep Sci 2012, 35(2), 231–238.

[105] Cheng L.-Y., Sun X.-M., Tang M.-Y., Jin R., Cui W.-G., Zhang Y.-G. An update review on recent skin fillers. Plast Aesthet Res 2016, 3, 92.

[106] Monheit G.D., Coleman K.M. Hyaluronic acid fillers. Dermatol Ther 2006, 19(3), 141–150.

[107] Kablik J., Monheit G., Yu L., Chang G., Gershkovich J. Comparative physical properties of hyaluronic acid dermal fillers. Dermatol Surg 2009, 35(Suppl 1), 302–312.

[108] Baumann L., Kaufman J., Saghari S. Collagen fillers. Dermatology Ther 2006, 19(3), 134–140.

[109] Beasley K., Weiss M., Weiss R. Hyaluronic acid fillers: a comprehensive review. Facial plast sur 2009, 25, 86–94.

[110] Parveen S., Misra R., Sahoo S.K. Nanoparticles: a boon to drug delivery, therapeutics, diagnostics and imaging. Nanomed Nanotechnol 2012, 8(2), 147–166.

[111] Kim J.-H., Park K., Nam H.Y., Lee S., Kim K., Kwon I.C. Polymers for bioimaging. Prog Polym Sci 2007, 32(8), 1031–1053.

[112] Saha T.K., Ichikawa H., Fukumori Y. Gadolinium diethylenetriaminopentaacetic acid-loaded chitosan microspheres for gadolinium neutron-capture therapy. Carbohyd Res 2006, 341(17), 2835–2841.

[113] Hallouard F., Anton N., Choquet P., Constantinesco A., Vandamme T. Iodinated blood pool contrast media for preclinical X-ray imaging applications – a review. Biomaterials 2010, 31(24), 6249–6268.

Emilie Prouvé, Gaétan Laroche, Marie-Christine Durrieu

4 Hydrogels for mesenchymal stem cell behavior study

Abstract: Mesenchymal stem cells (MSCs) are self-renewing, multipotent stem cells with the ability to differentiate into mesoderm-type cells, such as adipocytes, osteocytes, and chondrocytes. MSCs have also been reported to differentiate into other cell types such as neurons, smooth muscle cells, and hepatocytes in vitro. Consequently, they constitute an interesting candidate for tissue engineering and regenerative medicine purposes. However, the perfect control of MSC commitment toward a desired lineage has still not been achieved, which is an obstacle for their use in clinical applications. In this context, hydrogels have been identified as a promising tool to mimic the properties of the native extracellular matrix of cells and to investigate the response of cells to many different features. Indeed, hydrogel properties, such as topography, porosity, mechanical properties, and biomolecule presentation, are easily tunable and can lead to significantly different cell behavior in terms of cellular attachment, proliferation, and differentiation. In addition, hydrogels offer the possibility to encapsulate cells, and therefore compare cell response between two-dimensional environments, typically used in traditional cell culture experiments, and three-dimensional environments, more representative of the cell in in vivo environment. Therefore, this chapter aims at gathering the accumulated knowledge on the control of MSC behavior by using hydrogels as a cell culture material, which might help designing appropriate biomaterials for clinical applications.

4.1 Introduction

The interest in hydrogel materials is growing rapidly, considering their unique swelling properties, coupled with a high versatility and a high tunability of material's features, which opened the door to many applications such as disposable diapers, filters for water purification, separation materials for chromatography and electrophoresis, biosensors, cosmetic products, and drug delivery [1]. In particular, hydrogels have broad uses in biomedical research, including regenerative medicine and tissue engineering [2]. While most of the current understanding of cell processes is based on experiments performed on flat and stiff materials, such as polystyrene and glass, it became clear that these materials are not representative of the physiological environment of cells, and that culture systems that better mimic cells native in in vivo environments were needed [3]. In this context, hydrogels have proven useful in many cell culture applications, shedding light on

https://doi.org/10.1515/9781501519116-004

mechanisms regulating cell behavior and providing the appropriate conditions for the expansion and controlled differentiation of various cell types. Hydrogels constitute a powerful tool to mimic the properties of the native extracellular matrix (ECM) of cells and to investigate the response of cells to different features such as topography, porosity, mechanical properties, and biomolecule presentation. They are also gaining attention due to their ability to encapsulate cells [2], and therefore compare cell response between two-dimensional (2D) and three-dimensional (3D) environments.

Because of their self-renewal ability and potential to differentiate into multilineages, stem cells became a prominent subject in medical research for regenerative medicine and tissue engineering purposes [4]. Among stem cells, pluripotent stem cells, such as embryonic stem cells (ESCs) or induced pluripotent stem cells (iPSCs), are attractive because they can give rise to all the cell types in the body [4]. However, the use of ESCs in research and clinic is restricted due to ethical considerations. In addition, they can cause an immune response of the patient, as the cells come from another person, and can form tumors after they are cultivated in vitro and subsequently implanted in vivo [5]. iPSCs are an ethical alternative to ESCs as they can be obtained by reprogramming adult cells. Nevertheless, the cell reprogramming efficiency is quite low and the use of iPSCs in clinic is still limited because of the lack of knowledge regarding their manipulation and behavior [5, 6]. Considering the need for stable, safe, and highly accessible stem cell sources, adult stem cells, and particularly mesenchymal stem cells (MSCs), constitute interesting candidates. Indeed, they are self-renewable, multipotent, easily accessible, and can be expanded in vitro with high genomic stability and few ethical issues [4]. MSCs can be extracted from various tissues like bone marrow, adipose tissue, and dental pulp, and they are capable of in vivo differentiation into mesoderm-type cells such as osteoblasts, chondrocytes, and adipocytes, as well as other cell types such as neurons, smooth muscle cells, and hepatocytes when cultured in vitro [4]. Although MSCs exhibit several advantages, their clinical use is hindered by their tendency to uncontrollably proliferate and differentiate, which can lead to tumor formation [7]. Moreover, the perfect control of MSC differentiation toward a desired lineage has still not been achieved. Consequently, a better understanding of their biological behavior is required to provide tools to control their fate and allow their use in clinical applications.

Therefore, the objective of this chapter is to provide a broad overview of the use of hydrogels as cell culture substrate and the modulation of hydrogel properties to study and direct MSC adhesion, proliferation, and differentiation.

4.2 Impact of hydrogel properties on mesenchymal stem cell fate

4.2.1 Surface topography

In vivo, behavior of stem cells is controlled by the interplay of many different signals coming from cells surrounding microenvironment. This microenvironment is composed of chemical and mechanical cues, which will be reviewed in the following sections, as well as topographical cues at the micro- and nanoscale [8]. Surface topography is therefore considered as an interesting tool to alter cell growth, morphology, and differentiation. Indeed, micron-scale topographic features such as ridges, grooves, and pillars have been shown to influence cell spreading, migration, and differentiation [9–11]. Nanostructures including pits, pillars, and grooves have also been shown to elicit specific cell responses on several materials [12–14]. However, most of the studies performed so far have been conducted with solid substrates. Consequently, new methods have been developed for the fabrication of micro- and nanotopographies on the surface of hydrogel substrates, as hydrogels present the benefit of having similar features to ECM. The strategy consists in fabricating hydrogel substrates with micro- or nanotopographical patterns, on which the cells spread, elongate, and align through a phenomenon called contact guidance [15], which will determine cell shape and differentiation.

Li et al. developed different topographies on polyacrylamide hydrogels functionalized with type I collagen to promote cell adhesion, to study the effect of rat bone marrow MSC spreading, proliferation, and differentiation[16]. Substrate topography was set in square pillars or grooves, with three sets of dimensions. The width of the patterns and the distance between them were set to 5/15, 10/10, or 15/5 μm, for both squares and grooves. In addition, substrate stiffness was varied to 6 kPa (soft) and 47 kPa (stiff). This research team showed that both stiffness and dimension affect MSC proliferation, as stiffer and more unevenly dimensioned substrates (15/5 μm) lead to more cell growth. However, there was no difference in the multiplication rate between square pillars and grooves. Substrate topography is a key factor to regulate MSC morphology as a grooved topography promotes the directed alignment of cells and a reduced cell area compared to square pillars (Figure 4.1a and b). Finally, it has been shown that osteogenic differentiation was promoted on the stiff substrate with uneven square pillars (15/5 μm), while neurogenic differentiation was predominant on the soft substrate with evenly dimensioned grooves (10/10 μm), as grooved topography favors cell alignment and axon growth [16]. Similarly, Hu et al. molded rectangular microplates onto poly(2-hydroxyethyl methacrylate) (pHEMA) hydrogels [15]. The dimensions of the hydrogel microplates were 20 μm in height and 2 μm in width, and 10, 25, and 50 μm in length. The interplate spacing varied between 5 and 10 μm and the intercolumn spacing was 5 μm. It has been observed that human MSCs adhered and spread on a regular cell culture Petri dish but did not spread on a flat pHEMA hydrogel, due to the hydrophilicity of the pHEMA hydrogels and the lack of anchor point on the flat surface. However, cells adhered on microstructured pHEMA hydrogels and spread into elongated geometries in response to the topographical cues, as observed in the previous study, staying in between the parallel plates and conforming to the gaps (Figure 4.1c). Cell length increased as the interplate spacing decreased, as cells need to stretch more in the longitudinal direction to fit into a smaller space between the parallel plates. In addition, the cells tend to elongate more on longer plates. Finally, after 32 days of culture, the cells formed a dense and interconnected layer, with cells connecting in vertical direction through the intercolumn spacing. This is one of the advantages of having plate-patterned substrate over groove-patterned substrate, as it allows cell–cell connections and interactions (Figure 4.1c) [15]. Although many studies focused on polygonal patterns, as squares, grooves, and plates, some researchers also investigated the impact of circular patterns on cell behavior. For example, Yang et al. produced chitosan hydrogels with uniform microhills of 10 µm and dispersed microhills of 5–30 µm in diameter [17]. Spacing between microhills was determined as 4 µm for uniform hills and 14 µm for dispersed hills. They observed that rat MSCs on uniform hills were flat, polygonal, and well spread, while the majority of MSCs on dispersed hills were fusiform with cells that adhered to the hills, forming a bridge-like structure across the hills, or that fell between the hills. Therefore, cells on dispersed hills would spread and migrate along the spacing between microhills, leading to orientation by contact guidance. No difference in cell adhesion was found between flat and patterned hydrogels; however, MSCs proliferation was higher for dispersed hills followed by uniform hills and at last, flat hydrogels [17]. Although the spatially and dimensionally dispersed microhills were found to promote MSCs alignment and proliferation, it is not certain whether this effect is due to microhills size, shape, or to the spacing between the hills.

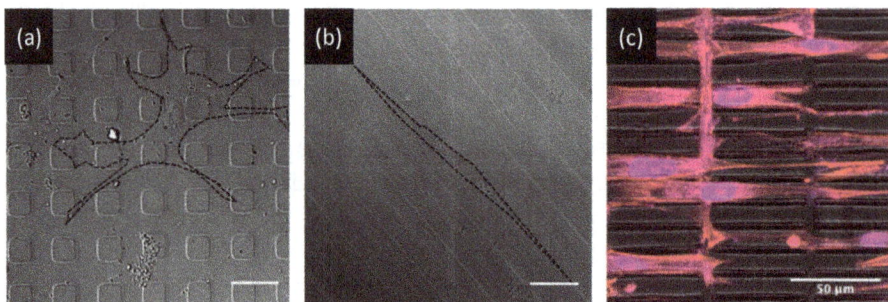

Figure 4.1: (a, b) Spreading of rat MSCs on a stiff (47 kPa) polyacrylamide hydrogel patterned with square pillars (a) or (b) grooves. The grooved topography promoted cell alignment by contact guidance and led to a reduced cell area compared with square pillars. Scale bars: 20 µm. Adapted from ref. [16]. (c) Human MSCs cultured on a microplate-patterned pHEMA hydrogel substrate. After 7 days of culture, cells elongated and aligned along the direction parallel to the plates and connected through the intercolumn spaces. Adapted from ref. [15].

Besides pillars and grooves, other hydrogel topographies consisting in wrinkles and cavities with various shapes have been studied. Poellmann et al. prepared polyacrylamide hydrogels with square or hexagonal cavities, varying the size of the cavities (from 3 to 20 μm) and the width of the borders separating the cavities (from 1 to 20 μm) [18]. They observed that the proportion of well-spread mouse bone marrow MSCs was higher for 10 and 15 μm borders than for 1 and 2 μm borders, with no effect of cavity shape and size. However, cell area was higher for square patterns with large borders (15–20 μm) than for hexagonal patterns or small borders (2–5 μm). Finally, square patterns induced cell alignment along the borders, with cells on smaller borders becoming more elongated, while they did not align on substrates with hexagonal cavities. This is explained by the fact that borders on square substrates are similar to continuous grooves, which is consistent with the results of the precedent studies, while borders between hexagonal cavities frequently change direction on a scale shorter than most cells [18]. Guvendiren et al. investigated the effects of pattern geometry and size on stem cell morphology and spreading by molding lamellar and hexagonal patterns with a periodicity (λ) of 50 or 100 μm and a height of 20 μm on pHEMA hydrogels [19]. For lamellar patterns with $\lambda = 50$ μm, most of the cells formed bridges between the patterns and spread randomly without recognizing the pattern, while for $\lambda = 100$ μm, the majority of cells aligned themselves along the patterns. For hexagonal patterns with $\lambda = 50$ μm, one-third of the cells attached inside the patterns and remained rounded, while for $\lambda = 100$ μm, the majority of the cells were found to be inside the patterns with a round morphology but a larger cell area. Lamellar patterns with $\lambda = 100$ μm and hexagonal patterns with $\lambda = 50$ μm were chosen to study human MSCs differentiation as they induced aligned cells which could stimulate osteogenesis and round cells which could favor adipogenesis, respectively. The differentiation on patterned surfaces was also compared with that observed on flat surfaces. As expected, MSC osteogenic differentiation was found to be upregulated for lamellar patterns, while adipogenic differentiation was higher for hexagonal patterns. In addition, osteogenesis was found only for cells that were elongated and aligned by taking the shape of the lamellar pattern, whereas only cells inside the hexagonal patterns which remained round with low spread area stained positive for adipogenesis [19]. These observations confirm that surface topography can direct cell shape, which in turn modulates cell differentiation. Finally, Randriantsilefisoa et al. synthesized polyethylene glycol (PEG) hydrogels presenting both micro- and nanowrinkles or creases to study the morphology of human MSCs [20]. On wrinkled hydrogels, cells developed a branched morphology with many cell interconnections and created additional wrinkles due to cell contractile forces. For hydrogels presenting creases, the cells were preferentially oriented and gathered along the creases, promoting a tissue-like formation by being close to one another [20].

Finally, less defined structures such as random surface roughness might also influence cell spreading and differentiation. For example, Hou et al. developed gelatin methacryloyl (GelMA) hydrogels with two different values of stiffness (4 kPa as soft gel and 31 kPa as stiff gel) and with varying surface roughness (R_s) from the nano- to microscale (from 200 nm to 1.1 μm) to study the effect on human MSCs adhesion, spreading, and osteogenic differentiation [21]. They found that cell-spreading area on soft hydrogels increased with increasing surface roughness, and that cells presented defined and aligned actin stress fibers for a roughness greater than 500 nm. On stiff hydrogels, the spreading of MSCs was slightly enhanced by increasing the roughness from 200 to 500 nm, but was restricted for higher roughness. In addition, focal adhesions were observed on soft gels from a roughness of 500 nm, whereas on the stiff gels, the maximum size of focal adhesions was obtained for an intermediate roughness of 500 nm. These results are correlated with MSC osteogenic differentiation (Figure 4.2) as the increase of surface roughness on soft hydrogels induced a higher expression of osteogenic markers. On the stiff hydrogels, the osteogenic differentiation increased linearly with the surface roughness, but decreased for roughness greater than 500 nm. Thus, the highest levels of osteogenic differentiation were observed for the conditions inducing the highest cytoskeletal and nuclear tension [21].

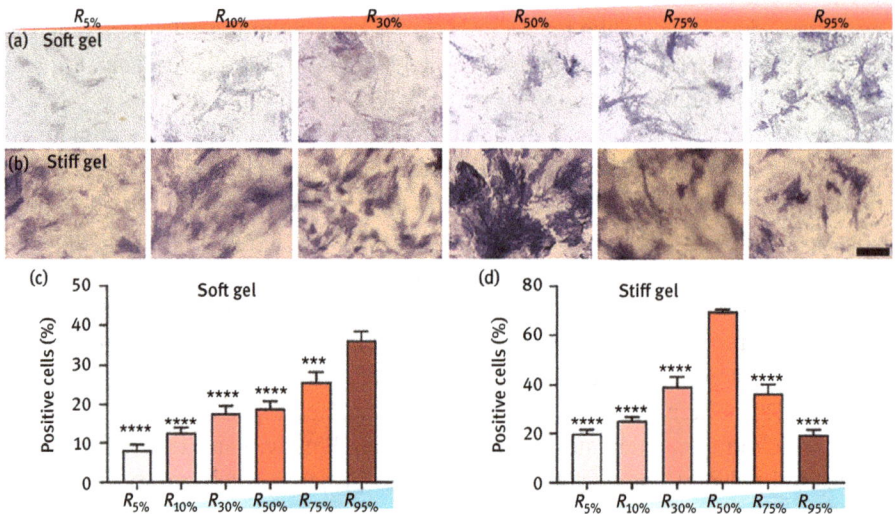

Figure 4.2: Osteogenic differentiation of human MSCs on roughness gradient hydrogels. Alkaline phosphatase staining of MSCs cultured on (a) soft (4 kPa) and (b) stiff (31 kPa) hydrogels in osteogenic-induced media for 7 days. Quantification of ALP positive cells on (c) soft and (d) stiff hydrogels. Scale bar: 200 μm. The designations $R_{5\%}$, $R_{10\%}$, $R_{30\%}$, $R_{50\%}$, $R_{75\%}$, and $R_{95\%}$ refer to roughness values (R_s) of 200, 300, 400, 500, 700, and 1,100 nm, respectively. On soft gels, MSC osteogenic differentiation increased with the roughness. On stiff gels, the differentiation was maximal for an intermediate roughness of 500 nm. From ref. [21].

4.2.2 Two-dimensional versus 3D culture systems

As previously mentioned, hydrogel materials are gaining popularity for cell culture applications, thanks to their ability to encapsulate cells [2], which allows cell response comparison between 2D and 3D environments. Expansion of MSCs is typically being conducted using traditional 2D-adherent culture conditions. This technique is relatively easy, but it has been demonstrated that cells produced in this manner eventually lose their stemness and differentiation potential, which is accompanied by replicative cell senescence and reduced paracrine capabilities [22]. Drawing inspiration from the native stem cell microenvironment, hydrogel platforms have been developed to drive stem cells fate by controlling parameters such as matrix mechanical properties, degradability, presence of cell-adhesive ligand, local microstructure, and cell–cell interactions [23, 24]. Consequently, the understanding of the differences in cell behavior as cell cultures are shifted from 2D surfaces to 3D substrates is highly needed. In general, 3D culture systems can be divided into two categories. The first category includes macroporous substrates that present interconnected pores with a pore size on the length scale of a single cell or greater (>10 µm) [25]. In these substrates, the pore size and architecture can influence cell behavior in terms of migration and differentiation [25, 26]. The second category comprises nonmacroporous substrates for which cells are fully encapsulated within the 3D substrate and are immobilized by contact with the substrate. Depending on the stiffness of the substrate and the ability of cells to degrade the hydrogel, the possibilities for cells to migrate or probe the surrounding environment will vary, which is likely to influence stem cell fate [25].

Nonmacroporous substrates have been proven useful for MSC chondrogenic differentiation as 3D cell encapsulation would allow rounded cell morphology and particular cell/cell or cell/matrix interactions, which are critical to obtain chondrocytes [27–29]. Merceron et al. used cellulose-based hydrogels to study the chondrogenic differentiation of human adipose tissue MSCs when cultured in 2D or 3D environments in vitro and in vivo [27]. After 3 weeks of culture in monolayers on top of the gel (2D) or encapsulated in gel pellets (3D) and in the presence of control or chondrogenic culture media, this team of scientists found that the expression level of four different chondrogenic markers was the highest for cells cultured in pellets and in the presence of chondrogenic medium. In addition, it has been shown that only 3D culture in the presence of chondrogenic medium supported the production of both glycosaminoglycan (GAG) and type II collagen, normally found in cartilage matrix. This study therefore demonstrated that 3D culture of MSCs was more suitable for chondrogenic differentiation in vitro, although a chondrogenic medium was required. Cells from the in vitro experiment were then collected and encapsulated in the hydrogels before being subcutaneously injected to mice, to ascertain whether cells were able to form cartilaginous tissue in vivo. After 5 weeks of implantation, cartilaginous tissue formation was achieved for MSCs treated with a chondrogenic medium in both 2D monolayer and 3D pellet cultures. While MSCs cultured in 2D chondrogenic medium failed to achieve chondrogenic differentiation in vitro, they formed a cartilaginous tissue to the same extent as MSCs precultured in 3D when they were transferred in 3D structures and implanted in vivo [27]. Although cells cultured in 2D and 3D led to the same result in vivo after 5 weeks of implantation, studying tissue formation at shorter time points might have revealed differences in tissue formation rate. Another study, conducted by Varghese et al., showed the benefit of 3D culture for MSC chondrogenic differentiation [30]. Nevertheless, this study highlighted that 3D culture alone was not sufficient to direct MSC differentiation toward the chondrogenic phenotype and that the physicochemical properties of the matrix were determinant to guide cell fate. Indeed, by encapsulating goat MSCs into PEG hydrogels or PEG hydrogels containing chondroitin sulfate (CS) moieties, it has been shown that PEG-CS hydrogels promoted self-aggregated cell clusters homogeneously distributed within the hydrogels and producing cartilaginous tissues, while no cell aggregation was observed for PEG hydrogels. Cell aggregation was correlated with chondrogenic differentiation as MSCs in PEG-CS hydrogels exhibited earlier activation and higher expression of chondrogenic markers compared with MSCs in PEG hydrogels. The authors explained these observations by the fact that CS moieties enhanced the aggregation of cells, possibly by interacting with various growth factors from the culture medium and enhancing their activity, leading to the formation of large cell clusters, which is recognized as a requirement for chondrogenesis. In addition, they showed that the immobilized CS segments of the scaffold underwent degradation in response to the cellular processes, which facilitated large cell cluster growth and matrix deposition [30]. Besides the presence of specific chemical moieties, the addition of ECM proteins or peptides into the 3D environment can also influence cell differentiation. For example, Jung et al. entrapped different proteins, such as type I collagen, laminin (LN), and fibronectin(FN) into PEG hydrogels, allowing cell encapsulation to study human MSC differentiation in 3D environments versus the situation where cells were cultured on 2D protein films [31]. For 2D culture, after 14 days, they found that myogenic differentiation was enhanced for a film of FN when compared with cells cultured on plastic dishes in the presence of differentiation medium, while none of the protein films was able to promote osteogenic, adipogenic, and chondrogenic differentiations to higher levels than the differentiation medium. For 3D matrices, after 28 days, hydrogels containing collagen were found to upregulate the expression of each lineage gene (myogenic, osteogenic, adipogenic, and chondrogenic) compared to hydrogels with LN and FN. In addition, ECM proteins in a 3D environment stimulated adipogenic and osteogenic differentiation at higher levels than in 2D culture [31]. However, even within a 3D environment, cell differentiation might not be homogeneous. Song et al. encapsulated human bone marrow MSCs within alginate hydrogels and directed their differentiation using osteogenic or adipogenic culture medium [32]. After 3 and 7 days, they observed homogeneous osteogenesis with similar amounts of calcium deposition through the hydrogel. Conversely, in the case of adipogenesis, more lipids were observed at the bottom of the gel than at the top [32]. Finally, while the above-mentioned studies have been conducted using covalently cross-linked hydrogels, the question that arises is whether the type of cross-linking (chemical or physical) could impact stem cell fate when encapsulated within the hydrogel. Huebsch et al. showed that within nondegradable, ionically cross-linked alginate hydrogels functionalized with RGD (Arg-Gly-Asp) peptides, encapsulated MSCs differentiation was dictated by matrix stiffness with osteogenic commitment occurring primarily at intermediate stiffness (11–30 kPa) and adipogenic lineage predominating in softer (2.5–5 kPa) microenvironments (Figure 4.3a) [33]. These observations were explained by two phenomena. First, it appeared that cell interaction with integrins was different depending on hydrogel stiffness, with, for example, a higher number of integrin α5-RGD bonds for a stiffness of 22 kPa. Second, it has been shown that cells used traction forces to mechanically reorganize the RGD peptides presented within these hydrogel matrices. These tractions forces are dependent on matrix stiffness as cells cultured on very compliant substrates cannot assemble their cytoskeleton and adhesion complexes while this is required to exert the so-called traction forces. On the contrary, on very rigid substrates, the cells cannot generate enough force to deform the matrix, which in turn will influence stem cell fate [33]. The possibility for cells to generate forces when encapsulated within these hydrogels is likely to be attributed to the physical cross-linking of alginate which allows matrix reorganization as physical cross-links can break and reform [28]. On the contrary, the study of Khetan et al. showed that human MSCs encapsulated within covalently cross-linked RGD-modified methacrylated hyaluronic acid (MeHA) hydrogels underwent almost exclusively adipogenesis relative to osteogenesis for all the tested stiffnesses from 4 to 91 kPa (Figure 4.3b and 4.3c) [34]. In addition, MSCs showed limited focal adhesion formation and unpolymerized actin in 3D matrices, while MSCs seeded on 2D MeHA gels of similar elastic modulus (25 kPa) exhibited focal adhesion and underwent primarily osteogenic differentiation. These results are correlated with the measurement of very minimal deformation of the surrounding gels by encapsulated MSCs in all formulations. To confirm that cell differentiation was related to the ability of cells to deform the matrix, MSCs were encapsulated in hydrogels with cleavable cross-links (CCs) allowing cell degradation or with permanent cross-links (PCs) inhibiting cell degradation. MSCs in CC gels developed a robust network of stress fibers and focal adhesions which were not observed in PC gels. In addition, MSCs spread within CC gels and deformed the surrounding matrix to a greater extent than in PC gels. When switching the culture medium to a mixed adipogenic/osteogenic medium, MSCs in CC and PC gels underwent primarily osteogenesis and adipogenesis, respectively (Figure 4.3d). Thus, for hydrogels of the same stiffness, osteogenesis was favored when cells were able to spread and pull on the surrounding matrix, and adipogenesis was favored when cells remained rounded and were unable to displace the surrounding matrix [34, 35]. Stem cell fate is therefore regulated by cell-generated tension that can be disabled by the presence of nondegradable covalent cross-links, which indicates that stem cells response to biophysical cues is highly dependent on the type of hydrogel used, as well as the dimensionality of the system.

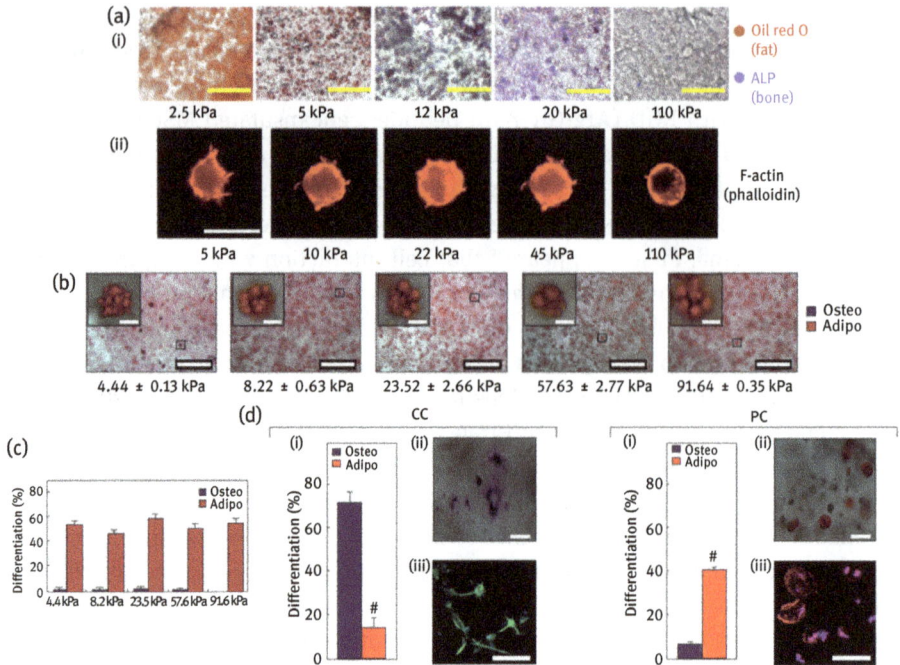

Figure 4.3: (a) (i) Staining of encapsulated mouse MSCs for ALP activity (fast blue: osteogenic marker, blue) and neutral lipid accumulation (oil red O: adipogenic marker, red) after 1 week of culture in the presence of combined osteogenic and adipogenic culture medium within RGD-modified alginate hydrogels with different stiffness. Adipogenesis was favored for soft substrates while osteogenesis occurred at intermediate stiffnesses. (ii) Actin staining of mouse MSCs 2 h after encapsulation into alginate matrices with varying stiffnesses. MSC differentiation was dictated by matrix stiffness irrespective of cell morphology as MSCs remained rounded independently of stiffness. Scale bars:
(i) 100 µm and (ii) 10 µm. Adapted from ref. [33]. (b) Representative bright-field images and (c) percentage differentiation of human MSCs within MeHA gels following 7 days of incubation in mixed osteogenic/adipogenic culture medium. Adipogenesis was predominant regardless of the stiffness. Scale bars: 100 and 5 µm (insets). (d) (i) Percentage of human MSC differentiation toward osteogenic or adipogenic lineages in gels with cleavable cross-links (CCs) or permanent cross-links (PCs) (#$p < 0.005$). (ii) Representative bright-field images of MSC staining for ALP (osteogenesis) and lipid droplets (adipogenesis) in CC or PC hydrogels. (iii) Representative immunocytochemistry for osteocalcin (OCN, osteogenesis, green) and fatty acid binding protein (FABP, adipogenesis, red) of MSCs in CC or PC hydrogels. Osteogenesis was favored for CC gels where cells were able to pull on the surrounding matrix, and adipogenesis was favored for PC gels where cells were unable to displace the surrounding matrix. Scale bars: (ii) 25 µm and (iii) 20 µm. Adapted from ref. [34].

In the case of macroporous substrates, the size and the architecture of pores are likely to affect MSCs differentiation. For example, Phadke et al. fabricatedPEG diacrylate-*co*-*N*-acryloyl 6-aminocaproic acid hydrogels with either randomly oriented pores of 50–60 μm (spongy gel) or lamellar pores of 100–150 μm (columnar gel) and showed that human MSCs cultured in spongy gels presented a more spread morphology when compared with cells seeded in columnar gels, which tended to form small cellular aggregates along the pore walls [36]. Although both pore architectures supported MSC osteogenic differentiation, cells in spongy gels exhibited significantly higher expression of several osteogenic markers and higher calcium deposition, suggesting that the spongy gels promoted faster osteogenic differentiation than the columnar gels. The enhanced osteogenic differentiation of MSCs in spongy gels could be explained by the highly interconnected porous network, which facilitated nutrients transport. The difference in tortuosity of the pores might also influence cell differentiation. Finally, the higher pore surface area in the spongy gels allowed for increased available area for cell spreading, leading to more spread cells which favor osteogenesis [36]. In another study, oligo(poly(ethylene glycol) fumarate) hydrogels were synthesized with a fixed pore size (50–100 μm) and with varying porosity of 0%, 20%, and 40%. It has been found that the gel with 40% porosity prolonged cell viability, which has been accounted for enhanced nutrient transfer. In addition, alkaline phosphatase (ALP) activity of rat bone marrow MSCs was higher for a porosity of 40% after 8 days, showing that greater porosity enhances osteogenic differentiation [37]. Using the same hydrogels but with constant porosity of 75% and pore size of 100, 300, and 400 μm, Dadstean et al. showed that rat bone marrow MSCs tended to aggregate on the edge and inside the pores of the interconnected porous network [38]. ALP activity and calcium deposition after 14 days were found to be significantly higher within porous hydrogels than on regular tissue culture plastic, but no difference was observed for the various pore sizes [38]. In addition to osteogenesis, hydrogel porosity has been found to influence chondrogenesis. Recently, Yang et al. developed covalently cross-linked type I collagen hydrogels with different pore architecture to study the impact on rat bone marrow MSCs chondrogenesis [39]. They fabricated hydrogels presenting a porous network with large and solid walls (the P group) with a pore size of 35 μm for P1 and 20 μm for P2, or hydrogels displaying a fibrous network with abundant micropores (the F group) with a median pore size of 0.7 μm for F1 and 0.3 μm for F2 (Figure 4.4a). MSCs were found to form clusters for all samples, but with larger clusters within the F2 gel. Cell proliferation was faster for the F group than for the P group, with F2 promoting the fastest proliferation. On the contrary, the cell area was higher for the P group, with a well-organized actin network after 7 days of culture, while cells presented a round morphology for the F group after 1 day and more spread cells but with dispersed actin cytoskeleton after 7 days (Figure 4.4b), indicating that chondrogenic phenotype was favored in the F group. Regarding cell differentiation, the expression of five different chondrogenic markers was higher in the F group after 7 days, with the highest gene expression observed for F2. Furthermore, GAGs and collagen II, produced in cartilage, were found in the F group but much less in the P group, suggesting a more chondrogenic cell type in the F group. Finally, cells in the P group presented calcium deposition indicating that more MSCs in the P group underwent osteogenic differentiation

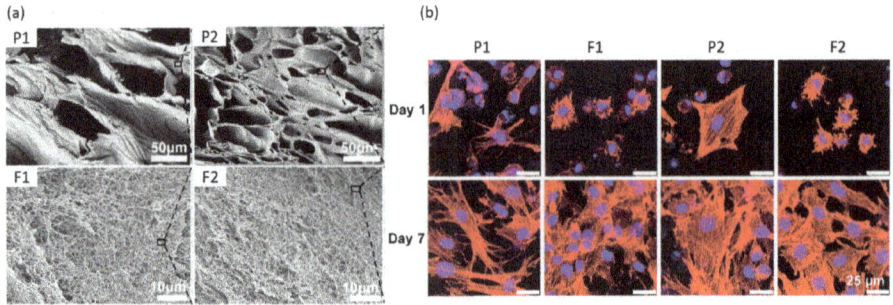

Figure 4.4: (a) SEM images of covalently cross-linked type I collagen hydrogels with different pore architecture. Hydrogels of the P group presented a porous network with large and solid walls (pore size of 35 μm for P1 and 20 μm for P2) and hydrogels from the F group displayed a fibrous network with abundant micropores (median pore size of 0.7 μm for F1 and 0.3 μm for F2). (b) Phalloidin (cytoskeleton in red)/DAPI (nuclei in blue) staining of rat bone marrow MSCs encapsulated in the hydrogels. Cell area was higher in hydrogels from the P group. Scale bars: 25 μm. Adapted from ref. [39].

than in the F group. Similar observations were made in vivo, leading to the conclusion that the F group hydrogels facilitated chondrogenesis. Similar to previous studies, the higher cell proliferation rate of the F group was explained by the abundant well-interconnected micropores facilitating proteins and cell metabolic wastes circulation. In addition, the dense porous network of F group hydrogels provided a confined space for cells which resulted in cells with decreased spreading area, dispersed actin cytoskeleton, and spherical morphology, directing cells toward the chondrocyte phenotype. On the contrary, the P group hydrogels had a larger porous network with relatively flat pore walls, providing flat surfaces for cells to spread, which was beneficial for osteogenic differentiation. This study also confirmed the results obtained by Khetan et al. [34] showing that MSCs tended to undergo osteogenic differentiation in degradable 3D hydrogels, as hydrogels from the P group exhibited faster degradation and higher osteogenic differentiation [39]. Finally, collagen–hyaluronic acid (HA) hydrogels with three distinct mean pore sizes (94, 130, and 300 μm) allowed an increase in rat bone marrow MSCs attachment by increasing the pore size [40]. In addition, for smaller pore size, cells adopted a flat morphology, whereas cells presented a rounded morphology for a larger pore size of 300 μm, which is correlated with a higher expression of chondrogenic markers for the largest pores. The chondrogenic differentiation occurring preferably for the largest pores can be explained by a better transport of chondrogenic factors and nutrients as well as a lower specific surface area which would result in lower cell adhesion ligand density, promoting rounded morphologies and therefore directing cells toward the chondrocyte phenotype [40].

4.2.3 Mechanical properties

The key to control and direct stem cell commitment into specific cell types required for regenerative medicine is thought to lie in mimicking the properties of the ECM of cells. This goes through surface conjugation with biomolecules, which will be discussed in the next section, and the control of the mechanical properties of the matrix. Indeed, considering that cells' mechanical microenvironments can be as physically diverse as brain, muscle, cartilage, or bone, it is thought that glass or plastic dishes, commonly used for standard in vitro cell culture, fail to provide proper environment for cell growth and differentiation[41]. Hence, over the past 15 years, a particular effort has been devoted to the development of hydrogels with tunable stiffness mimicking native tissues to study the impact of stiffness on cell adhesion, proliferation, and stem cells differentiation.

In particular, covalently cross-linked polyacrylamide hydrogels have been extensively used for such studies, as they offer the possibility to simply modulate hydrogel stiffness by varying their cross-linked content [42]. In pioneering work, Engler et al. synthesized polyacrylamide gels with varying stiffness and coated them with type I collagen to evaluate the effect of stiffness on human bone marrow MSC differentiation (Figure 4.5a) [43]. They demonstrated that MSCs grown on soft matrices mimicking brain tissue (0.1–1 kPa) tended to develop branched morphology and to express higher level of neurogenic differentiation markers, whereas on matrices with intermediate stiffness close to the modulus of muscle (8–17 kPa), cells exhibited a spindle-shaped typical of myoblasts, with greater expression of myogenic markers [43]. Finally, MSCs cultured on more rigid matrices mimic the stiffness of osteoid (25–40 kPa), a collagen matrix secreted by osteoblasts, became polygonal in shape, similar to osteoblasts and showed an upregulation of osteogenic markers [43]. These results suggest that optimizing the matrix mechanical properties could be a powerful tool to direct stem cells differentiation into a specific lineage. This hypothesis has been confirmed ever since by many other studies which are summarized in Table 4.1. Briefly, neurogenic [43–46] and adipogenic [33, 47–50] differentiation have been found to be predominant on soft matrices (from 0.1 to 5 kPa), myogenic commitment has been shown to be mostly encouraged for stiffnesses between 8 and 40 kPa [43, 45–47, 51, 52], while tenogenic differentiation was favored for stiffnesses between 30 and 50 kPa [53]. In addition, Yang et al. demonstrated that a soft matrix, with a stiffness close to that of bone marrow (2 kPa), would allow MSCs to maintain their stem cell phenotype [54]. Finally, the results obtained for chondrogenic and osteogenic differentiation were more heterogeneous, with chondrogenic differentiation reported both on soft matrices (0.5–1.5 kPa) [55, 56] and stiffer matrices (80 kPa) [46], and osteogenic differentiation mentioned for a wide range of stiffnesses going from 1.5 up to 190 kPa [33, 43, 44, 46, 48–51, 53–60], although it would be predominant between 20 and 90 kPa. Such inhomogeneity in the results could be explained by the use of a broad range of techniques to evaluate

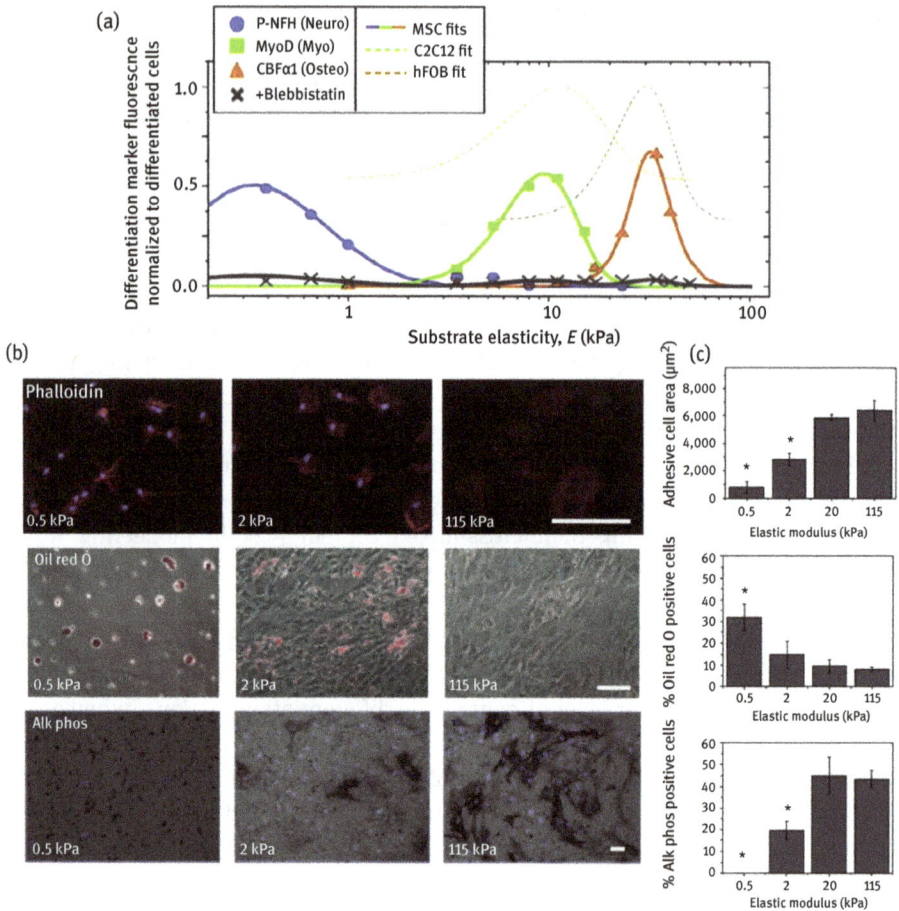

Figure 4.5: (a) Fluorescence intensity of human bone marrow MSCs differentiation markers versus substrate elasticity revealed maximal lineage specification at the stiffness typical of each tissue type. Average intensity was normalized to peak expression of control cells (murine myoblasts C2C12 or human osteoblasts hFOB). Blebbistatin blocked all marker expression in MSCs. Adapted from ref. [43]. (b) F-actin staining (phalloidin staining; red), oil red O, and alkaline phosphatase (Alk Phos) staining of human MSCs on polyacrylamide hydrogels. Nuclei were counterstained with DAPI (blue). Scale bars: 200 μm. (c) Quantification of MSCs spreading after 24 h and differentiation after 7 days (oil red O for adipogenic differentiation and alk phos for osteogenic differentiation) in culture on polyacrylamide hydrogels (*$p < 0.05$). Cell-spreading area and osteogenic differentiation increased with hydrogel stiffness but remained constant from a stiffness of 20 kPa, while adipogenic differentiation increased for decreasing stiffness. Adapted from ref. [47].

hydrogels stiffness, which complicates the comparison between the different studies. In addition, Trappmann et al. showed different results depending on the substrate used, as they obtained a stronger adipogenic differentiation of MSCs on soft polyacrylamide gels (0.5 kPa) and a stronger osteogenic differentiation for stiffer matrices

Table 4.1: Differentiation of mesenchymal stem cells seeded on hydrogels with varying stiffness.

Stem cell source	Hydrogel	Surface coating	Culture medium	Stiffness	Differentiation	Ref.
h-BM MSCs	Polyacrylamide	Type I collagen	Growth medium	0.1–1 kPa/8–17 kPa/ 25–40 kPa	Neurogenic 0.1–1 kPa	[43]
					Myogenic 8–17 kPa	
					Osteogenic 25–40 kPa	
h-BM MSCs	Polyacrylamide	Type I collagen	Growth medium	0.7–9–25–80 kPa	Myogenic 25–80 kPa	[51]
					Osteogenic 80 kPa	
h-BM MSCs	Polyacrylamide	Type I collagen	Growth medium	7–19–27–42 kPa	Osteogenic 42 kPa	[57]
h-BM MSCs	Polyacrylamide	Type I collagen	Growth medium	1–3–7–15 kPa	Adipogenic 1 kPa	[48]
					Myogenic 15 kPa	
h-MSCs	Polyacrylamide	Type I collagen	Mixed osteogenic and adipogenic medium	0.5–2–20–115 kPa	Adipogenic 0.5 kPa	[47]
					Osteogenic 20–115 kPa	
	PDMS	Type I collagen		0.1–40–800 kPa	No impact	
h-BM MSCs	Polyacrylamide	Type I collagen	Growth medium	Gradients 10–30 kPa/ 30–50 kPa/70–90 kPa	Tenogenic 30–50 kPa	[53]
					Osteogenic 70–90 kPa	
		Fibronectin			Osteogenic 30–50 kPa and 70–90 kPa	

(continued)

Table 4.1 (continued)

Stem cell source	Hydrogel	Surface coating	Culture medium	Stiffness	Differentiation	Ref.
h-BM MSCs	Polyacrylamide	Type I collagen	Osteogenic medium	1.5–26 kPa	Osteogenic 26 kPa	[58]
h-MSCs	Polyacrylamide	Type I collagen	Growth medium	1.6–40 kPa	Chondrogenic 1.6 kPa / Osteogenic 40 kPa	[55]
h-ASCs	Polyacrylamide	Type I collagen	Growth medium	4–13–30 kPa	Adipogenic 4 kPa / Osteogenic 30 kPa	[49]
h-BM MSCs	Polyacrylamide	Fibronectin	Growth medium	0.5–40 kPa	Neurogenic 0.5 kPa / Osteogenic 40 kPa	[44]
h-MSCs	Polyacrylamide	Fibronectin	Growth medium	15–37–50–63 kPa	Osteogenic 63 kPa	[59]
mouse-BM MSCs	Polyacrylamide	Fibronectin	Growth medium	15–50–63 kPa	Osteogenic 63 kPa	[60]
Mouse-MSCs / >h-MSCs	Alginate	RGD peptide	Mixed osteogenic and adipogenic medium	2.5–5–12–20–110 kPa	Adipogenic 2.5–5 kPa / Osteogenic 12–20 kPa	[33]
h-MSCs	Gelatin–hydroxyphenyl propionic acid	/	Growth medium	0.6–2.5–8–13 kPa	Neurogenic 0.6 kPa / Myogenic 13 kPa	[45]
Rat-BM MSCs	Collagen–glycosaminoglycan	/	Growth medium	0.5–1–1.5 kPa	Chondrogenic 0.5 kPa / Osteogenic 1.5 kPa	[56]

h-BM MSCs	Gelatin–hyaluronic acid	/	Adipogenic or osteogenic medium	0.15–1.5–4 kPa	Adipogenic 4 kPa; Osteogenic 4 kPa	[50]
h-MSCs	Silk fibroin	/	Growth medium	6–20–33–64 kPa	Myogenic 33 kPa	[52]
h-BM MSCs	Polyethylene glycol	RGD peptide	Mixed osteogenic and adipogenic medium	2–12 kPa	Stem cell 2 kPa; Osteogenic 12 kPa	[54]
h-BM MSCs	Polyvinyl alcohol–hyaluronic acid	/	Growth medium	Gradient 20–200 kPa	Neurogenic 20 kPa; Myogenic 40 kPa; Chondrogenic 80 kPa; Osteogenic 190 kPa	[46]

h, human; BM MSCs, bone marrow mesenchymal stem cells; ASCs, adipose stem cells; PDMS, Polydimethylsiloxane.

(20 and 115 kPa) (Figure 4.5b and c), but they did not observe any effect of the stiffness on MSC differentiation for PDMS substrates [47]. These results have been explained by the fact that cells do not directly pull the hydrogel matrix, but the covalently attached collagen on the surface of the gel. Consequently, the strains applied by cells are resisted by the attached collagen and the resistance is correlated to the number of anchorage points of collagen on the underlying matrix. Thus, by reducing the anchoring density of collagen on a gel with a stiffness of 20 kPa, cells did not exhibit the typical behavior for a stiffness of 20 kPa but behaved as if they were on a gel with a stiffness of 2 kPa [47]. These findings can explain the different behavior of cells between polyacrylamide hydrogels and PDMS, as well as the discrepancies observed over the literature. Finally, as most in vitro studies focused on cell state under static conditions at a particular time point, Lee et al. studied whether changing the biophysical aspects of the substrate could modulate the degree of MSCs lineage specification [44]. They explored MSC osteogenic and neurogenic differentiation on soft (0.5 kPa) or stiff (40 kPa) hydrogels followed by transfer of the cells to gels of the opposite stiffness. They observed that transferred MSCs, from soft to stiff gels, tended to decrease the expression of neurogenic markers and to increase the levels of osteogenic markers to levels that were comparable to cells that were cultured on the stiff gels alone. In the same way, transferred MSCs from stiff to soft gels showed a decreased expression of osteogenic markers and an increased expression of neurogenic markers as compared to cells maintained in culture on stiff substrates which mostly express osteogenic markers. Though the expression of osteogenic marker remained elevated compared to MSCs that were cultured on soft gels, indicating that transferring MSCs from stiff to soft substrates does not lead to a complete lineage reversal [44].

Besides cell differentiation, it has been shown that matrix stiffness has a great impact on cell adhesion, spreading, and proliferation. Many studies agree on the fact that cell area increases with matrix stiffness, with small and round cells with few stress fibers and focal adhesions on soft substrates, and highly spread cells with organized actin cytoskeleton and focal adhesions on stiff substrates [43, 48, 51, 54, 55, 61]. In addition, cell attachment and proliferation have been reported to be greater on stiffer matrices [45, 51, 52, 55, 60]. These observations would be explained by the fact that MSCs, as well as other cell types, can sense the rigidity of the substrate by exerting contractile forces, thanks to cell receptors as integrins, which can cluster into complexes known as focal adhesions that allow cell adhesion to the matrix. Focal adhesions can connect the substrate to cell cytoskeleton, providing physical links between mechanical environment and intracellular contractile architecture [62]. It is believed that cells on soft gels need to be less contractile than on stiff gels to adhere to the matrix. This is correlated with highly spread cells with organized actin cytoskeleton and stable focal adhesions on stiff gels, which maintain cell contractility, and more round cells with dynamic adhesions and no cytoskeleton organization on soft substrates [63]. Finally, as the cytoskeleton is associated with nuclear structures, the physical properties of cells are linked to gene expression [64]. In summary, matrix stiffness influences cell spreading and cytoskeleton organization which, in turn, drives cell differentiation through different pathways that have unfortunately not all been elucidated yet.

In this context, hydrogel substrates can also serve as a powerful tool for the measurement of traction forces exerted by cells on the ECM. The general method consists in cultivating cells on top of hydrogels containing thousands of fluorescent beads and following the displacements of the beads to calculate cell generated forces. For example, polyacrylamide [65, 66] or agarose [67] hydrogels have been used to study the distribution of forces exerted by fibroblasts or even metastatic breast adenocarcinoma cells in 2D culture. This method enabled to follow the dynamic mechanical interaction of the cells and their substrate and to characterize the location and magnitude of the traction forces. The same methodology was applied for 3D cell culture as it is more representative of the in vivo cell behavior. As such, Legant et al. encapsulated fibroblasts in 3D PEG hydrogels containing around 70,000 fluorescent beads and tracked the displacements of the beads in the vicinity of each cell to obtain a 3D map of cell-induced deformations [68]. Although these studies can provide useful results for the understanding of cell–matrix interactions, they have been conducted mainly with fully differentiated fibroblasts. Consequently, individual studies are required to investigate the behavior of different cell types, as the results obtained with fibroblasts are not directly transposable to other cells.

Although it is clear that matrix stiffness has a great impact on stem cell fate, it is now acknowledged that the mechanical properties of hydrogels and living tissues are not limited to elasticity, but include viscoelastic properties. Viscoelastic materials first resist cell-generated forces because of their elasticity, and then dissipate the applied forces over time due to their viscoelastic response, which can dramatically alter cell behavior [64]. However, only few studies investigated the impact of hydrogels viscoelastic properties on cell adhesion, proliferation, and stem cell differentiation. For example, Cameron et al. varied the formulation of polyacrylamide hydrogels to obtain matrices with a constant elastic modulus (4.7 kPa) and with varying viscous modulus (1, 10, and 130 Pa) [69]. They showed that human bone marrow MSCs' spread area and proliferation increase with the viscous modulus. In addition, adipogenic, myogenic, and osteogenic differentiation were enhanced for the highest viscous modulus in the presence of differentiation supplements [69]. Similarly, Charrier et al. developed polyacrylamide hydrogels with constant elastic modulus (5 kPa) and with varying viscous modulus (0, 200, and 500 Pa) by entrapping linear polyacrylamide into the hydrogel network [70]. They observed that mouse fibroblasts had overall smaller areas on viscoelastic gels than on purely elastic gels. In addition, they showed that the percentage of rat hepatic stellate cells undergoing myofibroblast differentiation decreased as the viscous modulus increased, and that differentiated cells had the ability to dedifferentiate on substrates with higher viscous modulus [70]. Chaudhuri et al. compared the effect of covalently cross-linked alginate hydrogels (elastic substrates) and chemically cross-linked alginate hydrogels (stress relaxing substrates) of varying elastic modulus (1.5, 3, and 9 kPa) on the spreading of human osteosarcoma cells and mouse fibroblasts [71]. They confirmed that the cell-spreading area increases with the hydrogel stiffness and they observed enhanced cell spreading on stress-relaxing substrates relative to elastic substrates, with greater cell focal adhesion formation and higher number of cells forming stress fibers [71]. Using similar hydrogels, Bauer et al. showed that mouse myoblasts had a significantly higher cell-spreading area as the elastic modulus was increased from 2.8 to 49.5 kPa on elastic gels, while for stress-relaxing gels, the cell-spreading area increased from 2.8 to 12.2 kPa, but did not further increase with higher modulus [72]. However, myoblast proliferation was higher on stress-relaxing substrates as compared to elastic substrates for all investigated elastic moduli [72]. In another study, Chaudhuri et al. developed alginate hydrogels using different molecular weight polymers in combination with different cross-linking densities of calcium to obtain matrices with constant elastic modulus but different relaxation time [73]. They evidenced that both spreading and proliferation of mouse fibroblasts increase with faster stress relaxation for a constant elastic modulus of 9 kPa. Moreover, they observed adipogenic differentiation of human MSCs for an elastic modulus of 9 kPa, with the level of adipogenesis decreasing in rapidly relaxing gels, and very low levels of osteogenic differentiation. In contrast, no adipogenic differentiation was observed for a higher initial elastic modulus of 17 kPa, while osteogenic differentiation was significantly enhanced in gels with faster stress relaxation (Figure 4.6) [73]. These studies highlight the importance of considering the effect of both elasticity and viscoelastic properties on cell behavior, as they can have great impact on cell adhesion, proliferation, and differentiation. In addition, these results seem to show that cell response to varying viscoelastic properties might depend on cell type. Consequently, more efforts should be dedicated to such studies in order to provide tools for the control of cell behavior for regenerative medicine purposes, as they are still lacking today. Finally, hydrogel materials offer many possibilities for mimicking the mechanical properties of the ECM and providing a suitable environment for controlled cell growth and fate.

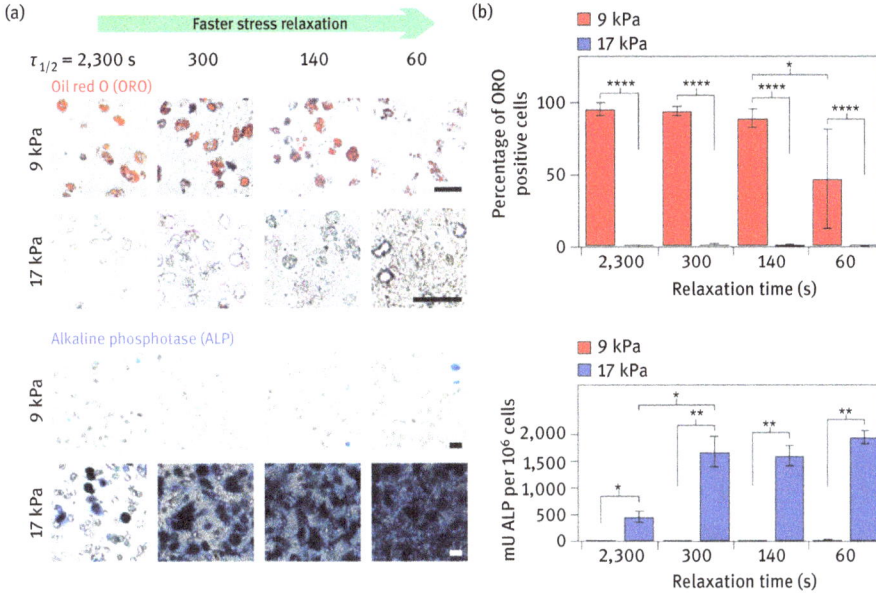

Figure 4.6: (a) Representative images of oil red O (ORO) staining (red), indicating adipogenic differentiation, and alkaline phosphatase (ALP) staining (blue), indicating early osteogenic differentiation, for human MSCs cultured in alginate gels of indicated modulus and timescale of stress relaxation for 7 days. Gels were functionalized with RGD peptide to promote cell adhesion. Scale bars: 25 µm. (b) Quantification of the percentage of cells staining positive for ORO, and a quantitative assay for ALP activity from lysates of cells in gels from the indicated conditions at 7 days in culture. *, **, and **** indicate $p < 0.05$, 0.01, and 0.0001, respectively. Adipogenesis occurred preferably for a soft gel with long relaxation times, while osteogenesis was favored on a stiffer gel with short relaxation times. Adapted from ref. [73].

4.2.4 Biofunctionalization

Over the past years, the conception of biomaterials has evolved from "inert" materials, which do not interact with the cells of the host organism, to sophisticated substrates with the capability to communicate with cells and direct their fate [74, 75]. Specifically, tuning the biochemical cues of the substrate has been identified as a promising way to direct cell fate. Many studies have therefore investigated the impact of several signaling molecules, such as growth factors, ECM proteins, and peptides, on cell behavior in terms of adhesion, spreading, and differentiation [74]. If these cues can be presented to cells in a soluble form, their immobilization on the material generally improves their stability and provides better control over their density and orientation [74, 75]. In the natural in vivo environment, it is acknowledged that cell adhesion and differentiation are partly mediated by the interaction with various proteins anchored in the ECM, which is mainly driven by integrin

receptors [75]. MSCs reside in a niche made up of neighboring cells and the ECM in-
fused with autocrine and paracrine soluble growth factors notably [76]. Cellular behavior depends on the abundance and distribution of bioactive factors in the ECM,
which undergoes remodeling [77] as a result of cells self-renewal and differentiation
[78]. For instance, during MSCs proliferation, the native ECM has a higher concentration in fibroblast growth factor-2 [79], while the ECM is richer in bone morphogenic
protein-2 (BMP-2) during osteogenic differentiation. Therefore, the classical approach
in materials biofunctionalization consists in mimicking the cellular microenvironment through interaction with growth factors, proteins, and peptides [75]. Growth
factors are soluble molecules that stimulate cell growth, differentiation, survival, and
tissue repair through specific binding of transmembrane receptors on target cells
[80]. The immobilization of growth factors and proteins to biomaterials prevents the
loss of bioactivity caused by their progressive release from the material in the soluble
form [80]. Nevertheless, the use of full-length proteins is limited by the complexity
and high costs of production and purification, as well as their lack of stability, as
they are very susceptible to changes of pH, temperature, and solvents-induced conformational changes [75]. In addition, proteins generally present different binding
sites that interact with various cell receptors and might trigger unwanted cellular responses [74]. The use of synthetic peptides, derived from a particular sequence of a
full-length protein, might circumvent these limitations as they can be produced in
large quantities with high purity and at low cost. Moreover, they are more stable to
pH and temperature changes, and they can be tuned to introduce anchoring units to
allow their binding to the surfaces without losing their biological activity [75].
Finally, besides biomolecules, some attention is also given to studying the effect of
different chemical functional groups on cell behavior.

If it is known that the ECM is composed of different proteins, it can also be considered at a smaller scale as it contains specific chemical functional groups depending on
the type of tissue. Indeed, carboxylic acid functionalities are typical of cartilage, as it is
rich in GAGs, while phosphates are highly present in mineralized tissue forming bone,
and hydrophobic functional groups are characteristics of adipose cells as they are rich
in lipids and release fatty acids [81]. For example, Wang et al. synthesized PEG hydrogels (PEG), PEG hydrogels containing phosphate groups (PhosPEG), or copolymer hydrogels of PEG and PhosPEG (PhosPEG-PEG), and showed that only PhosPEG–PEG
gels promoted osteonectin, collagen I, and ALP production of goat MSCs, indicating an
enhanced osteogenic differentiation[82]. In addition, calcium deposition was higher for
the copolymer hydrogels. The authors explained these differences by the degradation
rate of the various hydrogels, arguing that the presence of cleavable phosphoesters allowed PhosPEG-PEG hydrogels degradation which promoted mineralization by con-
verting the phosphoesters into insoluble calcium phosphate, which did not occur for
PEG gels. Finally, for PhosPEG gels, the degradation rate might be too high to promote
differentiation and matrix deposition [82]. Using polyacrylamide hydrogels modified by
plasma treatment to create surfaces with amino groups, carboxyl groups, or phosphate

groups, Lanniel et al. evidenced that human MSC differentiation was impacted by the combination of surface chemistry and matrix stiffness [83]. MSCs exhibited higher neurogenic differentiation for matrices with carboxyl groups and a stiffness of 6.5 kPa, while myogenic differentiation was enhanced for a stiffness of 10 kPa and carboxyl groups, and osteogenic differentiation was favored for a stiffness of 41 kPa and phosphate groups [83]. Similarly, Benoit et al. demonstrated that human MSCs cultured on PEG hydrogels carrying acid, phosphate, and t-butyl groups tended to differentiate into chondrocytes, osteoblasts, and adipocytes, respectively [81]. However, it is not clear whether these observations are due to direct interactions with the functional groups or with proteins that would be preferentially adsorbed on materials presenting specific chemical environments. Indeed, it has been demonstrated that human bone marrow MSCs had limited attachment on PEG hydrogels or phosphate-functionalized PEG hydrogels (PhosPEG) without serum in the culture medium, while PhosPEG gels promoted significantly higher cell attachment and spreading in the presence of serum compared to control PEG hydrogels, as PhosPEG hydrogels promoted higher protein adsorption [84]. In the same way, Ayala et al. controlled surface hydrophobicity of polyacrylamide-based hydrogels by varying the alkyl chain length of pendant side chains, terminated by a carboxylic acid function, to evaluate the impact on human MSCs behavior and protein adsorption [85]. It has been observed that cell adhesion and cell surface area increased progressively when increasing the length of the alkyl chain from one carbon (C1) to five carbons (C5), but decreased for six (C6), seven (C7), or ten carbons (C10). After 14 days of culture in osteogenic culture medium, several osteogenic markers were expressed for cells on C5 hydrogels while they were not detected for cells on C3 hydrogels (Figure 4.7). When cultured with myogenic culture medium, MSCs stained positive for myogenic markers on C5 hydrogels, while they underwent cell death on C3 hydrogels. These results were correlated with protein adsorption as these hydrogels were found to selectively adsorb FN and LN from the serum, with little proteins found on C1, C3, and C7 hydrogels, and abundant proteins adsorbed on C5 hydrogels. These observations were explained by the fact that the side chain must be sufficiently long to allow the terminal carboxyl groups to reach the binding sites on the proteins, explaining the small adsorption of proteins for hydrogels with short alkyl chains. However, when the side chains are too long, they become more hydrophobic and collapse onto the surface of the hydrogel, decreasing the accessibility of carboxyl groups for binding [85]. Therefore, MSCs response to surface chemistry is likely to be attributed to differences in protein adsorption rather than direct interaction with varying chemical groups (acid, phosphate, t-butyl groups, and alkyl chains).

The ECM is composed of different proteins, which interact with cell surface receptors as well as growth factors. If growth factors are usually solubilized in the culture medium for in vitro culture, the in vivo configuration is different as these molecules are sequestered in the ECM and interact with nearby cells [86]. In addition, as proteins and growth factors have short serum half-lives, their immobilization can increase the persistence of signaling and allows control of the delivered dose [86]. For

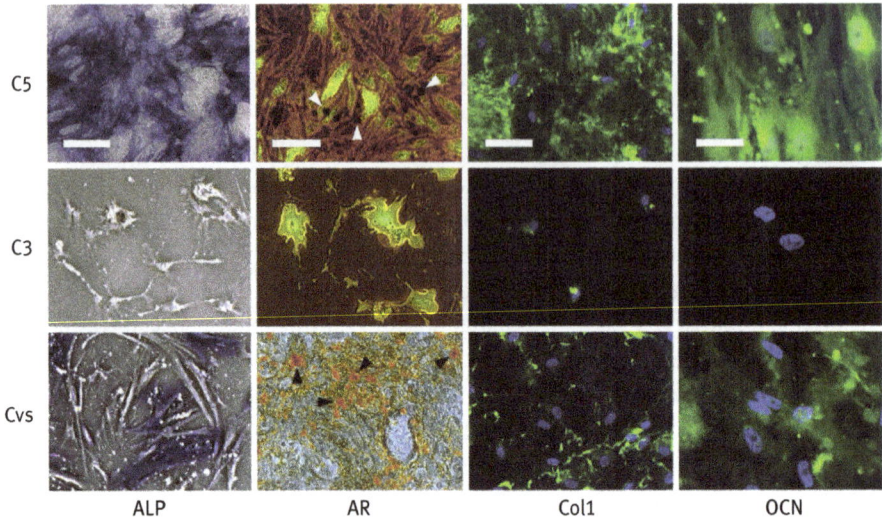

Figure 4.7: Osteogenic differentiation of human MSCs cultured on polyacrylamide-based hydrogels carrying surface alkyl chains with three (C3) or five (C5) carbons and on glass coverslips. Alkaline phosphatase (ALP) staining at 14 days in osteogenic medium. Alizarin red S (AR) and immunofluorescent (green) staining for collagen I (Col1) and osteocalcin (OCN) after 21 days. White (C5 hydrogels) and black (glass coverslips) arrowheads, in alizarin red S staining, point to calcium nodules. Scale bars: 200 μm (ALP), 100 μm (AR and Col1), and 50 μm (OCN). The C5 hydrogels led to higher osteogenesis. Adapted from ref. [85].

example, Rowlands et al. immobilized various ECM proteins (collagen I, collagen IV, FN, and LN) on polyacrylamide hydrogels with varying stiffness (0.7, 9, 25, and 80 kPa) and studied the impact on myogenic and osteogenic differentiation of human bone marrow MSCs (Figure 4.8) [51]. It has been found that MSCs osteogenic differentiation occurred predominantly on the stiffest gel (80 kPa) coated with collagen I, which may be attributed to the fact that this combination best mimics the natural microenvironment of bone. The osteogenic marker expression was very low on both collagen IV and LN-coated gels, regardless of gel stiffness, which suggests that cell differentiation is mediated by substrate stiffness in combination with the presented ECM molecule. For myogenic differentiation, the maximum expression of myogenic marker occurred on FN-coated gel with a stiffness of 25 kPa. Moreover, it has been observed that collagen I- and LN-coated gels showed the highest expression at a stiffness of 80 kPa, while cells cultured on FN- and collagen IV-coated gels had the greatest expression for a stiffness of 25 kPa, which confirms the interplay between stiffness and adhesive ligand presentation [51]. As different proteins and different combinations of proteins are susceptible to influence the cell fate, Dolatshahi-Pirouz et al. developed miniaturized human MSC-laden gelatin hydrogel constructs entrapping various proteins, including FN, LN, and OCN, to study the impact of these proteins on MSCs osteogenic differentiation [87]. They evidenced that constructs combining several proteins, and especially FN–OCN and LN–FN–OCN, resulted in higher ALP expression and calcium deposition, as compared to individual incorporation of ECM proteins. The authors explained that the functional properties of ECM proteins can be changed through protein–protein interactions, as it causes structural alterations that might expose hidden osteogenic regions on combined proteins, thereby enhancing the osteogenic differentiation of MSCs [87]. While functionalizing the hydrogel with ECM proteins is an approach to direct cell differentiation, the strategy of Benoit et al. consisted of immobilizing heparin on a PEG hydrogel, as heparin is capable of interacting with numerous proteins associated with osteogenic differentiation [88]. The authors showed that hydrogels functionalized with heparin led to higher expression of osteogenic markers of human MSCs as compared to functionalization with RGD peptide (a cell adhesion peptide present in several matrix proteins such as FN, vitronectin, osteopontin (OPN) and bone sialoprotein [89]) or with both heparin and RGD. These results were explained by a higher binding affinity of heparin with the proteins and growth factors from the culture medium, therefore promoting osteogenic differentiation [88]. The immobilization of growth factors such as transforming growth factor β (TGF-β), which regulates multiple biological processes including embryonic development, adult stem cell differentiation, immune regulation, wound healing, and inflammation [90], was also studied for its potential to direct cell fate. For example, Kopesky et al. observed that adsorbed TGF-β1 onto peptide or agarose hydrogels induced increased chondrogenesis of equine and bovine MSCs over MSCs cultured in TGF-β1-free conditions, and similar chondrogenesis to MSCs cultured with TGF-β1 in the culture medium [91]. In addition, agarose hydrogels with adsorbed TGF-β1 stimulated the production of full-length aggrecan, a constituent of cartilage, while the presence of TGF-β1 in the culture medium led to aggrecan cleavage. However, tethered TGF-β1 neither stimulated accumulation of cartilage components nor induced proliferation of MSCs encapsulated in hydrogels, which might be explained by a lower accessibility of TGF-β1 and by the accumulation of newly secreted matrix proteins that may block tethered TGF-β1 from cell receptors [91]. On the contrary, McCall et al. showed that tethered TGF-β on PEG hydrogels, with a concentration higher than 10 nmol/L, promoted human MSC chondrogenic differentiation to a similar extent than chondrogenic culture medium containing soluble TGF-β [86]. Finally, Ding et al. recently evaluated human MSCs differentiation into vascular smooth muscle cells (vSMCs) on PEG hydrogels functionalized with RGD peptides or with both RGD and TGF-β1 [92]. They demonstrated that all hydrogels led to higher gene expression compared to undifferentiated MSCs or primary vSMCs cultured on plastic, showing that the expression of vSMCs markers was predominantly regulated by soft matrix environment (stiffness of 5 kPa) regardless of the presence of soluble or tethered TGF-β1. However, cells on hydrogels with tethered TGF-β1 exhibited greater calcium intake and higher levels of intracellular calcium signaling, as well as greater cell contractility, which are typical features of vSMCs. These results suggested that tethered TGF-β1 is more performant in directing MSCs differentiation toward functional vSMCs compared to soluble TGF-β1 that is generally used [92].

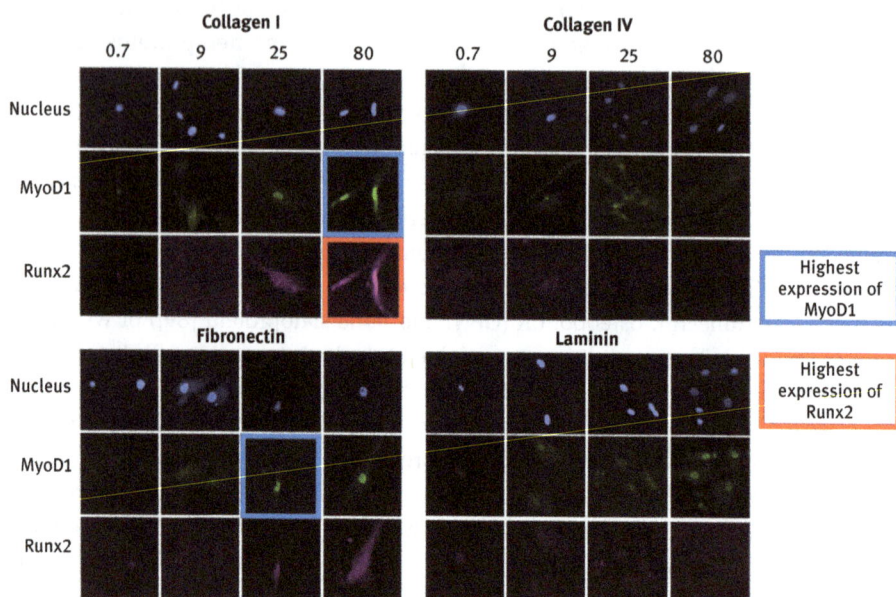

Figure 4.8: Representative images of human MSCs stained for MyoD1 (myogenic marker) and Runx2 (osteogenic marker) cultured on gels of various stiffness and protein coating. Runx2 expression was higher on stiff gels coated in collagen I, whereas MyoD1 was expressed in varying amounts on substrates with a stiffness higher than 9 kPa, regardless of protein coating. Adapted from ref. [51].

Synthetic peptides derived from ECM proteins represent a good alternative to the use of full-length proteins for materials biofunctionalization as they are readily available by synthetic methodologies, with tunable structure and high purity, and their use is exempt of immunogenic risks [75]. One of the first peptidic sequences used for biomaterial functionalization, which is still among the most widely used, is the cell-adhesion sequence Arg–Gly–Asp (RGD), isolated from FN [74]. Various studies have tethered RGD peptides on hydrogels, showing that peptides length, structure, and concentration could impact MSCs adhesion and differentiation. For example, Shin et al. showed that MSC adhesion was not enhanced on hydrogels surface if the molecular weight of the peptide was smaller than the molecular weight between hydrogel cross-links, as the peptide was buried inside the network and no receptor–ligand complex was formed [93]. The structure of the peptide is also likely to influence cell behavior as human MSC osteogenic differentiation was enhanced when encapsulated in alginate hydrogels presenting cyclic RGD peptides over linear RGD peptides, as cyclic peptides might promote more stable cell–ligand bonds [94]. Finally, peptide concentration is another parameter to consider to direct cell fate. It has been evidenced that the production of OCN and ALP of goat MSCs, encapsulated in PEG hydrogels functionalized with RGD peptides, increased with the RGD concentration (from 0.025 to 2.5 mmol/L), suggesting greater osteogenesis for the highest RGD concentration [95]. However, no influence of RGD concentration (from 0.1 to 2.5 mmol/L) was observed on the osteogenic differentiation of human ASCs cultured on 2D polyacrylamide gels [49]. Peptides derived from other proteins or designed to bind to specific molecules were also investigated. For example, MSCs encapsulated within alginate hydrogels functionalized with RGD or DGEA (Asp-Gly-Glu-Ala) peptides (derived from the $\alpha 2\beta 1$ integrin-binding domain of collagen I) expressed the highest level of osteogenic markers for hydrogels presenting the DGEA peptide [96]. Parmar et al. showed that the incorporation of HA-bound and CS-bound peptides, designed to bind HA and CS respectively, into collagen based hydrogels significantly enhanced the chondrogenic differentiation of human MSCs [97]. Specifically, the HA-bind hydrogels directed the highest increase in chondrogenic genes expression, leading to the greatest total collagen and GAGs accumulation as compared to CS-bound hydrogels, RGD hydrogels or hydrogels without surface conjugated peptides [97]. Similarly, HA hydrogels functionalized with a collagen mimetic peptide ($(GPO)_8$–CG–RGDS) promoted greater chondrogenic differentiation of rabbit bone marrow MSCs as compared to gels without surface conjugated peptides [98]. On the contrary, Connelly et al. reported that agarose hydrogels functionalized with three different peptides (a RGD peptide, a collagen mimetic peptide containing the GFOGER motif [Gly-Phe-hydroxy Pro-Gly-Glu-Arg], or a fragment of FN FnIII7-10) led to inhibition of cartilage matrix synthesis and chondrogenic gene expression of calf bone marrow MSCs [99]. These results support their previous observations that RGD peptides inhibited chondrogenic differentiation of calf MSCs encapsulated in alginate gels. In addition, this inhibition increased with increasing bulk densities of RGD in the gel [100]. Vega et al. observed the same trend of chondrogenic differentiation inhibition of human MSCs in HA hydrogels when increasing RGD concentration (from 0 to 5 mmol/L) [101]. However, they highlighted that a peptide containing the HAV (His-Ala-Val) motif extracted from N-cadherin, a transmembrane protein that mediates cell–cell adhesion and that is important for chondrogenesis, enhanced MSCs chondrogenic differentiation in a dose-dependent manner (Figure 4.9a, b, and c) [101]. These findings corroborate with the study of Kwon et al., as they evidenced that the inclusion of a HAV peptide into HA hydrogels enhanced human MSCs chondrogenesis in a dose-dependent manner. In addition, this effect was lost when the peptide was not permanently linked to the substrate [102]. Similarly, Bian et al. reported that conjugating a HAV peptide to HA hydrogels promoted chondrogenesis of MSCs and cartilage-specific matrix production as compared to unmodified gels or gels modified with a scrambled peptide [103]. Subcutaneous implantation of MSC-seeded hydrogels in mice also led to superior neocartilage formation in implants functionalized with the N-cadherin mimetic peptide compared with controls [103]. As N-cadherin would also be involved in the early stage of osteogenesis, Zhu et al. investigated the impact of a HAV peptide on human MSCs osteogenic differentiation [104]. It has been demonstrated that HA hydrogels functionalized with the combination of HAV and RGD peptides upregulated the expression of osteogenic markers, including type I collagen, OCN, ALP, and Runx2, and led to higher calcium content, as compared to hydrogels with RGD peptide or without peptide. The opposite effect was observed when the N-cadherin peptide was supplemented in the culture medium [104]. BMPs are a group of bioactive proteins that constitute important inducing factors during embryonic development and are closely related to osteoinduction [105]. Among them, BMP-2 plays a significant role in stimulating the differentiation of MSCs into osteoblasts by regulating the transcription of osteogenesis-related genes such as ALP, type-I collagen, OCN, and bone sialoprotein genes [105]. Consequently, several studies have used peptide sequences derived from the BMP-2 protein to promote MSCs osteogenic differentiation. Particularly, the peptide sequence corresponding to residues 73–92 of the knuckle epitope of recombinant human BMP-2 is mainly used as it would be implicated in osteogenic differentiation of bone marrow MSCs [106]. Alginate hydrogels functionalized with this peptide have been shown to promote higher MSCs ALP activity, as well as calcium and phosphate deposition, than hydrogels functionalized with RGD or DWIVA (Asp-Trp-Ile-Val-Ala, also extracted from BMP-2 protein) peptides, although the use of full-length BMP-2 protein led to the highest ALP activity, and calcium and phosphate deposition [107]. On polyacrylamide-based hydrogels, it has been found that surface modification with BMP-2 mimetic peptides engaged the commitment of human MSCs into osteogenic lineage regardless of the mechanical properties of the substrate (for stiffnesses of 15 and 47 kPa), except for very soft gels (stiffness of 0.76 kPa) [108]. In addition, synergistic effects have also been found between the BMP-2 mimetic peptide and other peptides. For example, He et al. showed that the functionalization of poly(lactide-co-ethylene oxide-co-fumarate) hydrogels with both RGD and BMP-2 peptides led to significantly higher ALP activity and calcium production than RGD conjugated or BMP grafted hydrogels [109]. Later, using the same hydrogels, this group confirmed that ALP activity of BMP-2 peptide-grafted gels was significantly higher than RGD grafted gels [110]. In addition, they found that gels grafted with a combination of RGD, BMP-2, and OPD (isolated from OPN) peptides presented the highest ALP activity and calcium deposition of rat MSCs. This combination has also proven to induce the strongest expression of vasculogenic markers (Figure 4.10) [110]. OPN is one of the noncollagenous proteins present in bone matrix. OPN is expressed in cells of the osteoblastic lineage and plays a critical role in the maintenance of bone [111]. Another peptide sequence isolated from OPN (ODP peptide) and containing the RGD motif has been grafted onto oligo(poly(ethylene glycol) fumarate) hydrogels to investigate the effect of the biomimetic surface on MSCs differentiation into osteoblasts [112, 113]. It has been found that the presence of signaling peptides (RGD and ODP) was favorable for MSCs osteogenic differentiation, although the differentiation and mineralization of the MSCs was not dependent on the peptide sequences used [112, 113]. Identifying proteins sequences responsible for a defined cell response is a key to control cell fate, as Lee et al. showed that the peptide corresponding to the residues 150–177 of human OPN allowed for higher ALP activity and mineralization of human MSCs when immobilized on alginate gels as compared to the peptide corresponding to the 53–80 sequence of OPN [114]. Finally, as the peptides sequences and the combination of peptides used are susceptible to influence cell fate, the development of microarrays might be an interesting tool for high-throughput screening of cellular behavior in multivariate microenvironments [115].

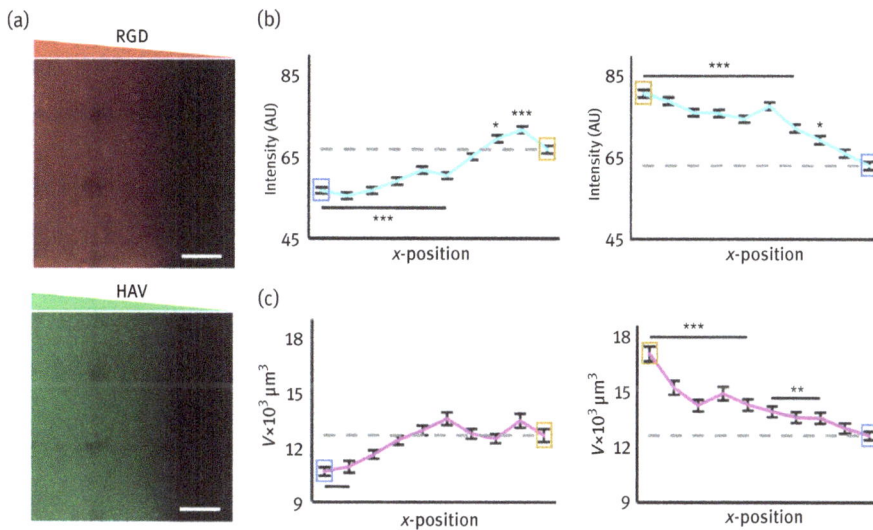

Figure 4.9: (a) Rhodamine-labeled RGD (GCGYGRGDSPG) or fluorescein-labeled HAV (HAVDIGGGC) peptide gradients on hyaluronic acid hydrogels. (b) Effects of HAV and RGD gradients on transcription factor Sox9 expression (chondrogenic marker). (c) Effects of HAV and RGD gradients on aggrecan synthesis. Generally, higher nuclear Sox9 and aggrecan content were observed with decreasing RGD and for increasing HAV. *, **, and *** indicate $p < 0.05$, 0.01, and 0.001, respectively, compared to lowest peptide region (gray dashed line). Adapted from ref. [101].

In addition to the composition and density of adhesion ligands on a substrate, the spatial distribution of these ligands has also been shown to influence MSCs behavior. For example, Kasten et al. designed adhesive lines of FN with varying width (between 10 and 80 μm) and spacings (between 5 and 20 μm) onto NCO-sP(EO-stat-PO) cell-repellent hydrogels [116]. This study revealed that human MSCs presented highly

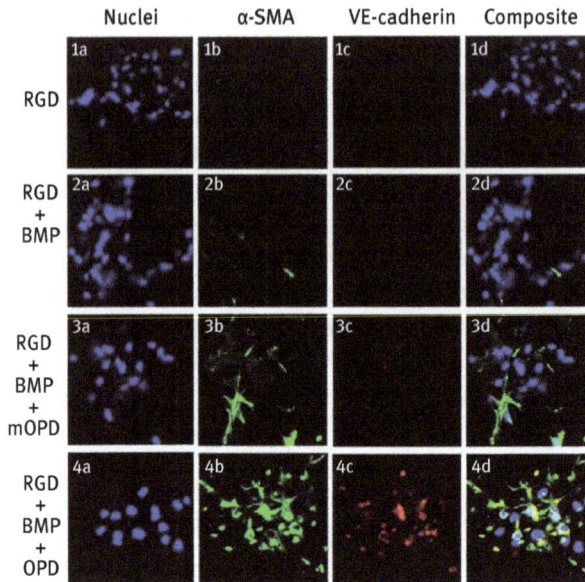

Figure 4.10: Effect of BMP-2 (BMP) and osteopontin (OPD) mimetic peptides grafting on RGD-conjugated hydrogels on the expression of vasculogenic markers of rat MSCs after 28 days of incubation. Hydrogels presenting the combination of RGD, BMP and OPD peptides promoted higher expression of vasculogenic markers (mOPD, mutant OPD peptide). From ref. [110].

aligned actin filaments with decreasing size of FN lines and a directed migration of cells was observed along the lines as opposed to homogeneous surface coating, with a higher migration rate with decreasing line width. These constructions enabled direct stem cells migration which would be important for tissue formation and regeneration. In addition, fabricating substrates with line patterns that allow cell bridging over nonadhesive gaps would mimic the ECM architecture as it is comprised of a fibrous network to which cells adhere and form bridges to cross the micron-sized gaps inside the filamentous network [116]. Patterning the adhesive ligands to form different shapes has also proven to be useful for directing MSCs differentiation. Microislands of RGD peptide, with different sizes (from 177 to 5,652 μm^2) and containing single cells, were made on PEG hydrogels to study the impact of cell size on MSCs differentiation (Figure 4.11a and b) [117]. It has been found that small cells preferred adipogenic commitment (from 177 to 1,413 μm^2), while large cells preferably underwent osteogenic differentiation (from 2,826 to 5,652 μm^2) [117]. Nanosized patterns might also influence cell fate as RGD nanopatterns with different spacing (37, 53, 77, 87, and 124 nm) made on PEG hydrogels showed a decreased cell density and spreading area with the increase of nanospacing, unexpectedly related to a higher MSCs osteogenic and adipogenic differentiations, under differentiation culture media, with the increase of RGD nanospacing [118]. These results indicate that micro- and

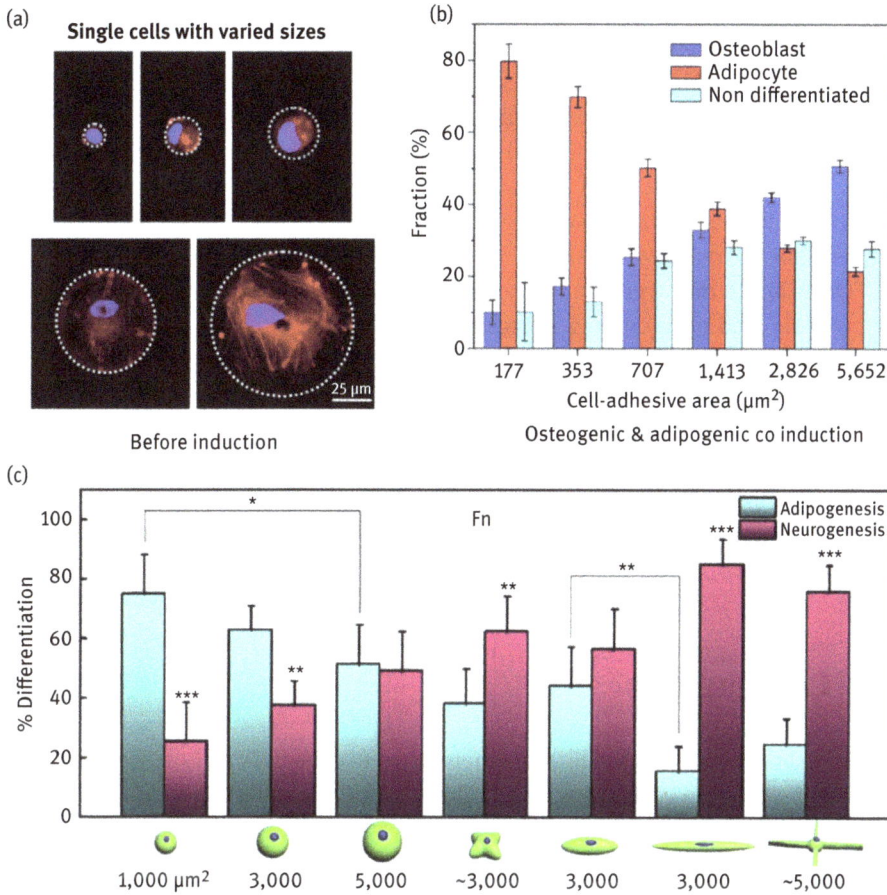

Figure 4.11: Differentiation of single MSCs on microislands of a series of sizes. (a) Fluorescent micrographs of single MSCs on microislands of varying size. Red: F-actin, blue: nuclei. (b) Percentages of adipogenesis, osteogenesis, and undifferentiation of single MSCs in mixed induction medium for 7 days. Small and large cells preferred to adipogenic and osteogenic commitments, respectively. Adapted from ref. [117]. (c) Percentage of human MSCs undergoing adipogenesis and neurogenesis on fibronectin patterns on polyacrylamide hydrogels (circular, oval, and star patterns). Smaller circular features promoted adipogenesis, while cells in anisotropic features preferred neurogenic differentiation. *, **, and *** indicate $p < 0.05$, 0.005, and 0.0005, respectively. Adapted from ref. [119].

nanopatterns both allow direct stem cells differentiation, although the effects are different. By varying the patterns shape of several proteins immobilized on polyacrylamide gels, Lee et al. demonstrated that smaller circular features promoted a higher expression of adipogenesis markers ($1{,}000\ \mu m^2 > 3{,}000\ \mu m^2 > 5{,}000\ \mu m^2$), while cells in anisotropic features such as four-branched stars and ovals preferred neurogenic differentiation (Figure 4.11c) [119]. In addition, MSCs cultured on FN tended to express elevated adipogenic markers while MSCs on collagen tended to express elevated neurogenic markers [119]. In another study, this research team cultured MSCs on unpatterned soft or stiff polyacrylamide gels for 10 days, and then transferred the cells to different stiffness substrates containing patterns of FN (circle, oval, star, or unpatterned; 5,000 μm^2) [44]. They highlighted that transferred cells (from soft to stiff) in oval and star shapes showed a higher expression of osteogenic markers compared to cells in other shapes, presumably because these shapes increase cytoskeleton tension which is known to promote osteogenesis. Then, MSCs that were transferred from stiff gels to oval and star shapes on soft gels showed an increased expression of neurogenic markers, demonstrating the importance of anisotropic geometries in guiding the extension of neuron-like processes [44]. These findings corroborated with a third study in which patterned MSCs in circular shapes displayed a disordered cytoskeleton without expression of osteogenic markers, while MSCs cultured in geometries that promoted an increased cytoskeletal tension (elongated oval shape and concave shape) showed a higher expression of osteogenic markers, particularly for a stiffness of 30 kPa [120].

4.3 Conclusions

It has become clear that biomimetic approaches to modulate cell–material interactions are essential to understand and further control cell behavior, which constitutes invaluable knowledge for the use of stem cells for regenerative medicine purposes. In this context, hydrogels have proven to be good candidates to mimic the in vivo ECM and to provide a suitable environment for cell growth, whether they are made of synthetic or natural polymers. Indeed, owing to their ability to absorb large amounts of fluids and allow nutrients and growth factors circulation, as well as their capacity to encapsulate cells, hydrogels have shown to maintain cell viability and promote cell growth. In addition, they can be tuned to mimic the properties of the ECM known to affect stem cell behavior and fate, such as physical properties including stiffness, viscoelasticity, pore size, and porosity. Spatial properties are also considered including dimensionality (2D or 3D) of the scaffold, degradation, and micro- and nanoscale topography of the surface with varying size, shape, and level of disorder. Finally, biochemical properties are also regarded through the presence of diverse proteins, growth factors or peptides and their spatial distribution. However, experimental studies often focus on some of these aspects, neglecting other important parameters, while the synergy between different cues from ECM is still little studied but is essential to provide long-term robust cell function. Furthermore, the variety of conditions used for cell culture might complicate the interpretations and comparisons between studies using different cell types (MSCs from bone marrow or adipose tissue, and from different animals) and different cell culture media (growth medium or differentiation media, addition of various supplements and growth factors). Although some clues and promising conditions have been identified to control stem cells fate, the differentiation of a population of cells is still not homogeneous and often requires the presence of differentiation culture medium. Nevertheless, the understanding of the influence of these physical, spatial, and biochemical properties on cell fate will enable the development of scaffolds with tailored properties adapted for specific regenerative medicine applications.

References

[1] Mahinroosta M., Farsangi Z.J., Allahverdi A., Shakoori Z. Hydrogels as intelligent materials : a brief review of synthesis, properties and applications. Mater Today Chem 2018, 8, 42–55.

[2] Tsou Y., Khoneisser J., Huang P., Xu X. Hydrogel as a bioactive material to regulate stem cell fate. Bioact Mater 2016, 1(1), 39–55.

[3] Caliari S.R., Burdick J.A. A practical guide to hydrogels for cell culture. Nat Methods 2018, 13 (5), 405–414.

[4] Ullah I., Subbarao R.B., Rho G.J. Human mesenchymal stem cells – current trends and future prospective. Biosci Rep 2015, 35(2), 1–18.

[5] Panek M., Marijanović I., Ivković A. Stem cells in bone regeneration. Period Biol 2015, 117(1), 177–184.

[6] Yousefi A., James P.F., Akbarzadeh R., Subramanian A., Flavin C., Oudadesse H. Prospect of stem cells in bone tissue engineering : a review. Stem Cells Int 2016, 2016.

[7] Van Pham P. Mesenchymal Stem Cells in Clinical Applications. Stem Cell. Springer, 2016.

[8] Dolatshahi-Pirouz A., Nikkhah M., Kolind K., Dokmeci M.R., Khademhosseini A. Micro- and nanoengineering approaches to control stem cell-biomaterial interactions. J Funct Biomater 2011, 2, 88–106.

[9] Watari S., et al. Modulation of osteogenic differentiation in hMSCs cells by submicron topographically-patterned ridges and grooves. Biomaterials 2012, 33(1), 128–136.

[10] Liguori G.R., et al. Directional topography influences adipose mesenchymal stromal cell plasticity : prospects for tissue engineering and fibrosis. Stem Cells Int 2019, 2019, 1–14.

[11] Cunha A., et al. Human mesenchymal stem cell behavior on femtosecond laser-textured Ti-6Al-4V surfaces. Nanomedicine 2015, 10, 725–739.

[12] Dalby M.J., et al. The control of human mesenchymal cell differentiation using nanoscale symmetry and disorder. Nat Mater 2007, 6(12), 997.

[13] Pedrosa C.R., et al. Controlled nanoscale topographies for osteogenic differentiation of mesenchymal stem cells. ACS Appl Mater Interfaces 2019, 11, 8858–8866.

[14] Wu Y., et al. The combined effect of substrate stiffness and surface topography on chondrogenicdifferentiation of Mesenchymal Stem Cells. Tissue Eng Part A 2017, 23, 43–54.

[15] Hu Y., You J., Aizenberg J. Micropatterned hydrogel surface with high-aspect-ratio features for cell guidance and tissue growth. ACS Appl Mater Interfaces 2016, 8, 21939–21945.

[16] Li Z., et al. Differential regulation of stiffness, topography, and dimension of substrates in rat mesenchymal stem cells. Biomaterials 2013, 34(31), 7616–7625.

[17] Yang J., Liu A., Zhou C. Proliferation of mesenchymal stem cell on chitosan films associated with convex micro- topography. J Biomater Sci Polym Ed 2011, 22 (October 2014), 919–929.

[18] Poellmann M.J., Harrell P.A., King W.P., Wagoner Johnson A.J. Geometric microenvironment directs cell morphology on topographically patterned hydrogel substrates. Acta Biomater 2010, 6(9), 3514–3523.

[19] Guvendiren M., Burdick J.A. The control of stem cell morphology and differentiation by hydrogel surface wrinkles. Biomaterials 2010, 31(25), 6511–6518.

[20] Randriantsilefisoa R., et al. Interaction of human mesenchymal stem cells with soft nanocomposite hydrogels based on polyethylene glycol and dendritic polyglycerol. Adv Funct Mater 2020, 30(1), 1905200.

[21] Hou Y., et al. Surface roughness and substrate stiffness synergize to drive cellular mechanoresponse. Nano Lett 2019, 20(December), 748–757.

[22] Bae Y., Kwon Y., Kim H., Lee S., Kim Y. Enhanced differentiation of mesenchymal stromal cells by three-dimensional culture and azacitidine. Blood Res 2017, 52, 18–24.

[23] Chaicharoenaudomrung N., Kunhorm P., Noisa P. Three-dimensional cell culture systems as an in vitro platform for cancer and stem cell modeling. World J Stem Cells 2019, 11(12), 1065–1084.

[24] Madl C., Heilshorn S. Engineering hydrogel microenvironments to recapitulate the stem cell niche. Annu Rev Biomed Eng 2018, 20, 21–47.

[25] Haugh M.G., Heilshorn S.C. Integrating concepts of material mechanics, ligand chemistry, dimensionality and degradation to control differentiation of mesenchymal stem cells. Curr Opin Solid State Mater Sci 2016, 20(4), 171–179.

[26] Harley B.A.C., Kim H., Zaman M.H., Yannas I.V., Lauffenburger D.A., Gibson L.J. Microarchitecture of three-dimensional scaffolds influences cell migration behavior via junction interactions. Biophys J 2008, 95(8), 4013–4024.

[27] Merceron C., et al. The effect of two- and three-dimensional cell culture on the chondrogenic potential of human adipose-derived mesenchymal stem cells after subcutaneous transplantation with an injectable hydrogel. Cell Transplant 2011, 20, 1575–1588.

[28] Zhang L., Yuan T., Guo L., Zhang X. An in vitro study of collagen hydrogel to induce the chondrogenic differentiation of mesenchymal stem cells. J Biomed Mater Res – Part A 2012, 100(March), 2717–2725.

[29] Zheng L., et al. Chondrogenic differentiation of mesenchymal stem cells induced by collagen-based hydrogel : an in vivo study. J Biomed Mater Res – Part A 2010, 93, 783–792.

[30] Varghese S., Hwang N.S., Canver A.C., Theprungsirikul P., Lin D.W., Elisseeff J. Chondroitin sulfate based niches for chondrogenic differentiation of mesenchymal stem cells. Matrix Biol 2008, 27, 12–21.

[31] Jung J.P., Bache-Wiig M.K., Provenzano P.P., Ogle B.M. Heterogeneous differentiation of human mesenchymal stem cells in 3D extracellular matrix composites. Biores Open Access 2016, 5, 37–48.

[32] Song J.H., Lee S.-M., Yoo K.-H. Label-free and real-time monitoring of human mesenchymal stem cell differentiation in 2D and 3D cell culture systems using impedance cell sensors. RSC Adv 2018, 8, 31246–31254.

[33] Huebsch N., et al. Harnessing traction-mediated manipulation of the cell/matrix interface to control stem-cell fate. Nat Mater 2010, 9(6), 518–526.

[34] Khetan S., Guvendiren M., Legant W.R., Cohen D.M., Chen C.S., Burdick J.A. Degradation-mediated cellular traction directs stem cell fate in covalently crosslinked three-dimensional hydrogels. Nat Mater 2013, 12(5), 1–8.

[35] Khetan S., Katz J.S., Burdick J.A. Sequential crosslinking to control cellular spreading in 3-dimensional hydrogels. Soft Matter 2009, 5(8), 1601–1606.

[36] Phadke A., et al. Effect of scaffold microarchitecture on osteogenic differentiation of human Mesenchymal Stem Cells. Eur Cell Mater 2015, 25, 114–129.

[37] Wang L., Lu S., Lam J., Kasper F.K., Mikos A.G. Fabrication of cell-laden macroporous biodegradable hydrogels with tunable porosities and pore sizes. Tissue Eng Part C 2015, 21 (3), 263–273.

[38] Dadsetan M., et al. Effect of hydrogel porosity on marrow stromal cell phenotypic expression. Biomaterials 2009, 29(14), 2193–2202.

[39] Yang J., et al. Influence of hydrogel network microstructures on mesenchymal stem cell chondrogenesis in vitro and in vivo. Acta Biomater 2019, 91(May), 159–172.

[40] Amos M., Gleeson J.P., O'Brien F.J. Scaffold mean pore size influences mesenchymal stem cell chondrogenic differentiation and matrix deposition. Tissue Eng Part A 2014, 00(00), 1–12.

[41] Lv H., et al. Biomaterial stiffness determines stem cell fate. Life Sci 2017, 178, 42–48.

[42] Denisin A.K., Pruitt B.L. Tuning the range of polyacrylamide gel stiffness for mechanobiology applications. ACS Appl Mater Interfaces 2016, 8(34), 21893–21902.

[43] Engler A.J., Sen S., Sweeney H.L., Discher D.E. matrix elasticity directs stem cell lineage specification. Cell 2006, 126(4), 677–689.

[44] Lee J., Abdeen A.A., Kilian K.A. Rewiring mesenchymal stem cell lineage specification by switching the biophysical microenvironment. Sci Rep 2014, 4, 5188.

[45] Wang L., Boulaire J., Chan P.P.Y., Chung J.E., Kurisawa M. The role of stiffness of gelatin-hydroxyphenylpropionic acid hydrogels formed by enzyme-mediated crosslinking on the differentiation of human mesenchymal stem cell. Biomaterials 2010, 31, 8608–8616.

[46] Oh S.H., An D.B., Kim T.H., Lee J.H. Wide-range stiffness gradient PVA/HA hydrogel to investigate stem cell differentiation behavior. Acta Biomater 2016, 35, 23–31.

[47] Trappmann B., et al. Extracellular-matrix tethering regulates stem-cell fate. Nat Mater 2012, 11(8), 642–649.

[48] Park J.S., et al. The effect of matrix stiffness on the differentiation of mesenchymal stem cells in response to TGF- b. Biomaterials 2011, 32, 3921–3930.

[49] Wen J.H., et al. Interplay of matrix stiffness and protein tethering in stem cell differentiation. Nat Mater 2014, 13(10), 979–987.

[50] Zhao W., Li X., Liu X., Zhang N., Wen X. Effects of substrate stiffness on adipogenic and osteogenic differentiation of human mesenchymal stem cells. Mater Sci Eng C 2014, 40, 316–323.

[51] Rowlands A.S., George P.A., Cooper-White J.J. Directing osteogenic and myogenic differentiation of MSCs: interplay of stiffness and adhesive ligand presentation. AJP Cell Physiol 2008, 295(4), C1037–C1044.

[52] Floren M., Bonani W., Dharmarajan A., Motta A., Migliaresi C., Tan W. Human mesenchymal stem cells cultured on silk hydrogels with variable stiffness and growth factor differentiate into mature smooth muscle cell phenotype. Acta Biomater 2015, 31, 156–166.

[53] Sharma R.I., Snedeker J.G. Paracrine Interactions between mesenchymal stem cells affect substrate driven differentiation toward tendon and bone phenotypes. PLoS One 2012, 7, 1–11.

[54] Yang C., et al. Spatially patterned matrix elasticity directs stem cell fate. Proc Natl Acad Sci 2016, 31, 4439–4445.

[55] Xue R., Li J.Y.-S., Yeh Y., Yang L., Chien S. Effects of matrix elasticity and cell density on human mesenchymal stem cells differentiation. J Orthop Res 2013, 31(9), 1360–1365.

[56] Murphy C.M., Matsiko A., Haugh M.G., Gleeson J.P., O'Brien F.J. Mesenchymal stem cell fate is regulated by the composition and mechanical properties of collagen –glycosaminoglycan scaffolds. J Mech Behav Biomed Mater 2012, 11, 53–62.

[57] Shih Y.R.V., Tseng K.F., Lai H.Y., Lin C.H., Lee O.K. Matrix stiffness regulation of integrin-mediated mechanotransduction during osteogenic differentiation of human mesenchymal stem cells. J Bone Miner Res 2011, 26(4), 730–738.

[58] Witkowska-Zimny M., Walenko K., Wrobel E., Mrowka P., Mikulska A., Przybylski J. Effect of substrate stiffness on the osteogenic differentiation of bone marrow stem cells and bone-derived cells. Cell Biol Int 2013, 37(6), 608–616.

[59] Sun M., et al. Extracellular matrix stiffness controls osteogenic differentiation of mesenchymal stem cells mediated by integrin α 5. Stem Cell Res Ther 2018, 9, 52.

[60] Sun M., et al. Effects of matrix stiffness on the morphology, adhesion, proliferation and osteogenic differentiation of mesenchymal stem cells. Int J Med Sci 2018, 15, 257–268.

[61] Sunyer R., Jin A.J., Nossal R., Sackett D.L. Fabrication of hydrogels with steep stiffness gradients for studying cell mechanical response. PLoS One 2012, 7(10), 1–9.

[62] Hao J., et al. Mechanobiology of mesenchymal stem cells: a new perspective into the mechanically induced MSC fate. Acta Biomater 2015, 20, 1–9.

[63] Discher D.E., Janmey P., Wang Y. Tissue cells feel and respond to the stiffness of their substrate. Science (80-) 2005, 310(5751), 1139–1143.

[64] Vining K.H., Mooney D.J. Mechanical forces direct stem cell behaviour in development and regeneration. Nat Rev Mol Cell Biol 2017, 18, 728–742.

[65] Lo C., Wang H., Dembo M., Wang Y. Cell movement is guided by the rigidity of the substrate. Biophys J 2000, 79, 144–152.

[66] Franck C., Maskarinec S.A., Tirrell D.A., Ravichandran G. Three-dimensional traction force microscopy : a new tool for quantifying cell-matrix interactions. PLoS One 2011, 6(3).

[67] Toyjanova J., Hannen E., Bar-kochba E., Darling E.M., Henann D.L., Franck C. 3D viscoelastic traction force microscopy. Soft Matter 2014, 10(3), 8095–8106.

[68] Legant W.R., Miller J.S., Blakely B.L., Cohen D.M., Genin G.M., Chen C.S. Measurement of mechanical tractions exerted by cells within three-dimensional matrices. Nat Methods 2010, 7(12), 969–971.

[69] Cameron A.R., Frith J.E., Cooper-White J.J. The influence of substrate creep on mesenchymal stem cell behaviour and phenotype. Biomaterials 2011, 32(26), 5979–5993.

[70] Charrier E.E., Pogoda K., Wells R.G., Janmey P.A. Control of cell morphology and differentiation by substrates with independently tunable elasticity and viscous dissipation. Nat Commun 2018, 9(2018), 1–13.

[71] Chaudhuri O., et al. Substrate stress relaxation regulates cell spreading. Nat Commun 2015, 6(6634), 1–7.

[72] Bauer A., et al. Hydrogel substrate stress-relaxation proliferation of mouse myoblasts. Acta Biomater 2017, 62, 82–90.

[73] Chaudhuri O., et al. Hydrogels with tunable stress relaxation regulate stem cell fate and activity. Nat Mater 2016, 15(3), 326–334.

[74] Maia F.R., Bidarra S.J., Granja P.L., Barrias C.C. Functionalization of biomaterials with small osteoinductive moieties. Acta Biomater 2013, 9(11), 8773–8789.

[75] Mas-Moruno C. Surface functionalization of Biomaterials for Bone Tissue Regeneration and Repair. Amsterdam, Netherlands, Elsevier Ltd., 2018.

[76] Discher D., Mooney D., Zandstra P. Growth factors, matrices, and forces combine and control stem cells. Science (80-) 2009, 324, 1673–1677.

[77] James A. Review of signaling pathways governing MSC osteogenic and adipogenic differentiation. Science 2013, 1.

[78] Keung A., Kumar S., Schaffer D. Presentation counts: microenvironmental regulation of stem cells by biophysical and material cues. Annu Rev Cell Dev Biol 2010, 26, 533–556.

[79] Tsutsumi S., et al. Retention of multilineage differentiation potential of mesenchymal cells during proliferation in response to FGF. Biochem Biophys Res Commun 2001, 288, 413–419.

[80] Akter F. Principles of Tissue Engineering. Amsterdam, Netherlands, Elsevier Inc., 2016.

[81]   Benoit D.S.W., Schwartz M.P., Durney A.R., Anseth K.S. Small functional groups for controlled differentiation of hydrogel-encapsulated human mesenchymal stem cells. Nat Mater 2008, 7(October), 816–823.
[82]   Wang D., Williams C.G., Yang F., Cher N., Lee H., Elisseeff J.H. Bioresponsive phosphoester hydrogels for bone tissue engineering. Tissue Eng 2005, 11(1), 201–213.
[83]   Lanniel M., Huq E., Allen S., Buttery L., Williams P.M., Alexander M.R. Substrate induced differentiation of human mesenchymal stem cells on hydrogels with modified surface chemistry and controlled modulus. Soft Matter 2011, 7(14), 6501–6514.
[84]   Gandavarapu N.R., Mariner P.D., Schwartz M.P., Anseth K.S. Extracellular matrix protein adsorption to phosphate-functionalized gels from serum promotes osteogenic differentiation of human mesenchymal stem cells. Acta Biomater 2013, 9(1), 4525–4534.
[85]   Ayala R., et al. Engineering the cell-material interface for controlling stem cell adhesion, migration, and differentiation. Biomaterials 2011, 32(15), 3700–3711.
[86]   Mccall J.D., Luoma J.E., Anseth K.S. Covalently tethered transforming growth factor beta in PEG hydrogels promotes chondrogenic differentiation of encapsulated human mesenchymal stem cells. Drug Deliv Transl Res 2012, 2, 305–312.
[87]   Dolatshahi-Pirouz A., et al. A combinatorial cell-laden gel microarray for inducing osteogenic differentiation of human mesenchymal stem cells. Sci Rep 2014, 4, 1–9.
[88]   Benoit D.S.W., Durney A.R., Anseth K.S. The effect of heparin-functionalized PEG hydrogels on three-dimensional human mesenchymal stem cell osteogenic differentiation. Biomaterials 2007, 28, 66–77.
[89]   Petrie T.A., Garcia A.J. Extracellular matrix-derived ligand for selective integrin binding to control cell function. In Biological Interactions on Materials Surfaces. New York, Springer., 2009, 133–156.
[90]   Lifshitz V., Frenkel D. TGF-β. Handbook of Biologically Active Peptides: neurotrophic Peptides 2013, 1647–1653.
[91]   Kopesky P.W., et al. Controlled delivery of transforming growth factor b1 by self-assembling peptide hydrogels induces chondrogenesis of bone marrow stromal cells and modulates smad2/3 signaling. Tissue Eng Part A 2011, 17, 83–92.
[92]   Ding Y., Johnson R., Sharma S., Ding X., Bryant S.J., Tan W. Tethering transforming growth factor β1 to soft hydrogels guides vascular smooth muscle commitment from human mesenchymal stem cells. Acta Biomater 2020, 105, 68–77.
[93]   Shin H., Jo S., Mikos A.G., Modulation of marrow stromal osteoblast adhesion on biomimetic oligo [poly(ethylene glycol) fumarate] hydrogels modified with Arg-Gly-Asp peptides and a poly (ethylene glycol) spacer, in 28th Annual Meeting of the Society for Biomaterials, 2002.
[94]   Hsiong S.X., Boontheekul T., Huebsch N., Mooney D.J. Cyclic arginine-glycine-aspartate peptides enhance three-dimensional stem cell osteogenic differentiation. Tissue Eng Part A 2009, 15(2), 263–272.
[95]   Yang F., Williams C.G., Wang D., Lee H., Manson P.N., Elisseeff J. The effect of incorporating RGD adhesive peptide in polyethylene glycol diacrylate hydrogel on osteogenesis of bone marrow stromal cells. Biomaterials 2005, 26, 5991–5998.
[96]   Mehta M., Madl C.M., Lee S., Duda G.N., Mooney D.J. The collagen I mimetic peptide DGEA enhances an osteogenic phenotype in mesenchymal stem cells when presented from cell-encapsulating hydrogels. J Biomed Mater Res – Part A 2015, 103, 3516–3525.
[97]   Parmar P.A., et al. Collagen-mimetic peptide-modifiable hydrogels for articular cartilage regeneration. Biomaterials 2015, 54, 213–225.
[98]   Ren Y., Zhang H., Qin W., Du B., Liu L., Yang J. A collagen mimetic peptide-modified hyaluronic acid hydrogel system with enzymatically mediated degradation for mesenchymal stem cell differentiation. Mater Sci Eng C 2020, 108, 110276.

[99]   Connelly J.T., Petrie T.A., García A.J., Levenston M.E. Fibronectin- and collagen-mimetic
       ligands regulate BMSC chondrogenesis in 3D hydrogels. Eur Cell Mater 2016, 22, 168–177.
[100] Connelly J.T., Garcia A.J., Levenston M.E. Inhibition of in vitro chondrogenesis in RGD -
       modified three-dimensional alginate gels. Biomaterials 2007, 28, 1071–1083.
[101] Vega S.L., et al. Combinatorial hydrogels with biochemical gradients for screening 3D
       cellular microenvironments. Nat Commun 2018, 9, 1–10.
[102] Kwon M.Y., Vega S.L., Gramlich W.M., Kim M., Mauck R.L., Burdick J.A. Dose and timing of
       N-Cadherin mimetic peptides regulate MSC chondrogenesis within hydrogels. Adv Healthc
       Mater 2018, 7, 1701199.
[103] Bian L., Guvendiren M., Mauck R.L., Burdick J.A. Hydrogels that mimic developmentally
       relevant matrix and N-cadherin interactions enhance MSC chondrogenesis. Proc Natl Acad
       Sci 2013, 110, 10117–10122.
[104] Zhu M., Lin S., Sun Y., Feng Q., Li G., Bian L. Hydrogels functionalized with N-cadherin
       mimetic peptide enhance osteogenesis of hMSCs by emulating the osteogenic niche.
       Biomaterials 2016, 77, 44–52.
[105] Yang J., et al. Bone morphogenetic proteins : relationship between molecular structure and
       their osteogenic activity. Food Sci Hum Wellness 3 2014, 3(2014), 127–135.
[106] Saito A., Suzuki Y., Ogata S., Ohtsuki C., Tanihara M. Accelerated bone repair with the use of
       a synthetic BMP-2- derived peptide and bone-marrow stromal cells. J Biomed Mater Res –
       Part A 2004, 72, 77–82.
[107] Madl C.M., Mehta M., Duda G.N., Heilshorn S.C., Mooney D.J. Presentation of BMP-2
       mimicking peptides in 3D hydrogels directs cell fate commitment in osteoblasts and
       mesenchymal stem cells. Biomacromolecules 2014, 15, 445–455.
[108] Zouani O.F., Kalisky J., Ibarboure E., Durrieu M.C. Effect of BMP-2 from matrices of different
       stiffnesses for the modulation of stem cell fate. Biomaterials 2013, 34(9), 2157–2166.
[109] He X., Ma J., Jabbari E. Effect of grafting RGD and BMP-2 protein-derived peptides to a
       hydrogel substrate on osteogenic differentiation of marrow stromal cells. Langmuir 2008, 24
       (16), 12508–12516.
[110] He X., Yang X., Jabbari E. Combined effect of osteopontin and BMP-2 derived peptides grafted
       to an adhesive hydrogel on osteogenic and vasculogenic differentiation of marrow stromal
       cells. Langmuir 2012, 28, 5387–5397.
[111] Noda M., Denhardt D.T. Osteopontin. Principles of Bone Biology 2008, 351–366.
[112] Shin H., Zygourakis K., Farach-carson M.C., Yaszemski M.J., Mikos A.G. Modulation of
       differentiation and mineralization of marrow stromal cells cultured on biomimetic hydrogels
       modified with Arg-Gly-Asp containing peptides . J Biomed Mater Res – Part A 2004, 69,
       535–543.
[113] Shin H., et al. Osteogenic differentiation of rat bone marrow stromal cells cultured on
       Arg–Gly–Asp modified hydrogels without dexamethasone and b-glycerol phosphate.
       Biomaterials 2005, 26, 3645–3654.
[114] Lee J., et al. Injectable gel with synthetic collagen -binding peptide for enhanced
       osteogenesis in vitro and in vivo. Biochem Biophys Res Commun 2007, 357, 68–74.
[115] Sharma S., Floren M., Ding Y., Stenmark K.R., Tan W., Bryant S.J. A photoclickable peptide
       microarray platform for facile and rapid screening of 3-D tissue microenvironments.
       Biomaterials 2017, 143, 17–28.
[116] Kasten A., et al. Guidance of mesenchymal stem cells on fibronectin structured hydrogel
       films. PLoS One 2014, 9(10), 1–10.
[117] Peng R., Yao X., Cao B., Tang J., Ding J. The effect of culture conditions on the adipogenic and
       osteogenic inductions of mesenchymal stem cells on micropatterned surfaces. Biomaterials
       2012, 33(26), 6008–6019.

[118] Wang X., Yan C., Ye K., He Y., Li Z., Ding J. Effect of RGD nanospacing on differentiation of stem cells. Biomaterials 2013, 34(12), 2865–2874.

[119] Lee J., Abdeen A.A., Zhang D., Kilian K.A. Directing stem cell fate on hydrogel substrates by controlling cell geometry, matrix mechanics and adhesion ligand composition. Biomaterials 2013, 34(33), 8140–8148.

[120] Lee J., Abdeen A.A., Huang T.H., Kilian K.A. Controlling cell geometry on substrates of variable stiffness can tune the degree of osteogenesis in human mesenchymal stem cells. J Mech Behav Biomed Mater 2014, 38, 209–218.

Didier Snoeck

5 Superabsorbent polymers as a solution for various problems in construction engineering

Abstract: One of the most-used man-made materials is concrete, a mixture of cement, sand, aggregates, water, and admixtures. It can be seen everywhere: in tunnels, bridges, and high-rise buildings. Ever since concrete was rediscovered two centuries ago, it has been studied in detail in order to optimize the material and to solve its inherent problems. Most people know that concrete is gray, hard, and strong, and expect it to last decades and even centuries. Unfortunately, this is not always the case. Concrete is a material which can cope with high compressive forces but when it is subjected to tensile forces, it may crack. This cracking is based on several environmental and loading conditions, but the fact that concrete is prone to crack is a big issue. When cracking occurs and potentially harmful substances enter the interior of concrete, the concrete matrix may be damaged and even be destroyed. That is the reason why a lot of maintenance and repair works are due in order to increase the durability and lifetime of structures in civil engineering. One way of dealing with these issues is the modification of the material itself, making it less prone to cracking and the durability-related consequences. An example is the use of reinforcements, coping with the tensile forces when concrete cracks. Cracks are not harmful but intruding substances may trigger the corrosion of the iron rebars leading to structural failure, which is again unwanted. In consequence, along the history, different materials were investigated and added to concrete to solve the previous adverted problems. So, why not try adding the white powder superabsorbent polymers in the cementitious material in order to solve these issues?

5.1 Introduction

Almost two decades ago [1, 2], superabsorbent polymers (SAPs) found their way into concrete technology and ever since were investigated in detail [3–7]. The typical SAPs used have the feature to absorb up to several hundred times their own weight in aqueous solutions due to osmotic pressure, resulting in the formation of a swollen hydrogel. Typically, SAPs made by polyacrylates (using acrylic acid and acryl amide as main chains) or natural polymers (such as alginates) [3, 6] are added dry to a cementitious mixture and used to solve various problems, which will be discussed in the following sections. SAPs could have different shapes. The use of SAPs with irregular shapes obtained by bulk polymerization or spheres obtained by suspension polymerization are the most studied and used in this area. Moreover,

https://doi.org/10.1515/9781501519116-005

grape-shaped and fiber types are studied as well [8]. As SAPs swell, they will absorb part of the mixing water causing the loss in workability of the fresh cementitious mixture. It is commonly accepted to add additional water to compensate for this loss in workability [3, 9–11]. How much they swell and which tests could be performed to quantify the swelling ability are found in literature [5, 9, 12]. The main tests to determine the swelling characteristics upon application in concrete are the filtration test and tea bag test. For the filtration test, a predetermined amount of SAPs is weighed. Next, an amount of liquid (water, seawater, sulfate solution, cement filtrate solution, etc.) is added. The SAPs are then able to absorb the liquid. At a certain time interval, the whole is filtered and the amount of liquid not absorbed by the SAPs is determined. Using the weight difference of the added liquid and the remaining liquid over the initial mass of the SAPs, the absorption capacity is determined. When looking at the tea bag method, a predetermined amount of dry SAPs is added in a sealed tea bag, which is submerged in the testing liquid. By weighing the tea bag at regular time intervals, the amount of liquid absorbed can be determined and used to calculate the absorption capacity. In case the swelling time is needed, a vortex test can be used [5, 13]. This test uses a magnetic stirrer in a cup and a predetermined amount of SAPs to exactly absorb the amount of testing fluid. By measuring the time for a vortex created by magnetic stirring to disappear, the swelling time can be estimated. The swelling characteristics, next to the absorption and desorption kinetics, are important parameters for the application in a cementitious material. Furthermore, the stability of the polymer over time must be used during the service life of a concrete structure [14].

In the following sections, various problems occurring in cementitious materials are addressed. The first one is the control of rheology and the second is the occurrence of shrinkage cracks. Third, concrete is susceptible to freeze-thawing and scaling in general. Fourth, it is permeable and, at last, it cracks. In all these cases, the addition of SAPs to the cementitious mixing could solve these issues and their application will be discussed in detail.

5.2 Changing the rheology by absorbing mixing water

As SAPs are mixed in, they will absorb a fluid they may encounter. In this way, they may change the interaction in a cementitious system, due to partial absorption of the mixing water. The SAPs cause a change in rheology due to their swelling ability [1, 2, 15–17]. This swelling capacity is dependent on the composition of the mixture and the composition of the SAPs. The resulting osmotic pressure is dependent on the external fluid composition and the chemical structure, length, and cross-linking degree of the SAP [14, 16–18]. Due to the uptake of mixing water by the SAPs, the workability decreases as less water is available in the cementitious matrix itself. Later on, this water will be released for mitigating shrinkage, as will be discussed in the next section.

In addition, the workability can be controlled, which is interesting for 3D printing technologies using cementitious materials [19, 20]. Linked to shrinkage- and durability-related issues, the 3D printing of cementitious materials still faces some problems such as autogenous shrinkage, which may be counteracted using SAPs. Continuous layering of printed specimens with SAPs (Figure 5.1) already proved to be successful [19]. This is a recent field of study and many parameters may be investigated.

Figure 5.1: Three-dimensional printing of cementitious materials with SAPs by continuous layering.

5.3 Shrinkage mitigation by internal curing

The first application of SAPs in cementitious materials was the use of their swelling behavior and water-retention capability to induce internal curing in order to counteract occurring shrinkage cracks [1, 2, 21–25]. Concrete, composed of the hydrating cement and water, possesses the problem of shrinking when water is receding, especially in systems with low water-to-binder ratios. The latter systems, such as (ultra)high-performance concrete [26, 27], are used more often due to their denser matrix, high strength, and featured workability. In the following sections, the different forms of shrinkage, the role that they play in a cementitious material and the use of SAPs for its mitigation are discussed.

5.3.1 Plastic shrinkage mitigation

Shrinkage occurs from the start, due to drying and harsh environmental conditions. This so-called plastic shrinkage may cause cracking during the first few hours after casting [28]. Due to the quick drying and bleeding of concrete, high capillary pressures are exerted in the cementitious matrix. If deformations are restrained, for example, due to the formwork or the reinforcements, the concrete will show cracking [29]. That is why concrete is postcured when possible. External curing, for example, by (fog-)spraying or covering with a plastic sheet, is not as sufficient as an internal curing approach. By using SAPs, internal curing is possible. The approach is still new and a lot of research needs to be performed in order to optimize the mitigation of plastic shrinkage. Currently, the SAPs were only able to partially mitigate this type of shrinkage. It was found that by adding 0.6 m% of SAPs (versus cement weight), the capillary pressures and plastic deformations were reduced, while the settlement deformations increased [27]. A study using nuclear magnetic resonance (NMR) to monitor the water kinetics with SAPs when plastic shrinkage conditions were imposed showed that 0.22 m% of SAPs were able to reduce plastic settlement and reduced plastic shrinkage cracking but were not able to completely mitigate it [30]. The SAPs were able to protect the cement paste internally from the harsh ambient drying conditions and were able to sustain the internal relative humidity (RH) below 5 mm of the cementitious surface. Results by the RILEM TC-260 RSC support these findings in an international round robin test [31]. In this study, concrete slabs were subjected to harsh drying and wind conditions in order to stimulate plastic shrinkage. SAPs were added during mixing in amounts of 0.15 and 0.3 m%. The results on plastic shrinkage mitigation seemed to be dependent on the type of SAP and whether they possess retentive properties. The water can be released early to influence the bleeding characteristics or can be released after setting to aid with internal curing. Both effects seem to play a role. No additional information on the SAPs was disclosed, and the main difference seemed to be the cross-linking degree [31, 32].

5.3.2 Autogenous shrinkage mitigation

Other forms of shrinkage include autogenous shrinkage as a result of cement hydration, thermally induced shrinkage, drying shrinkage due to the loss of water to the surroundings, and shrinkage due to carbonation. As cement reacts with water, hydration products precipitate in the water-filled spaces between the solid particles in the cementitious material. The water in the remaining small capillaries forms menisci and exerts hydrostatic tension forces. These capillary forces reduce the distance between the solid particles, leading to autogenous shrinkage. Chemical and

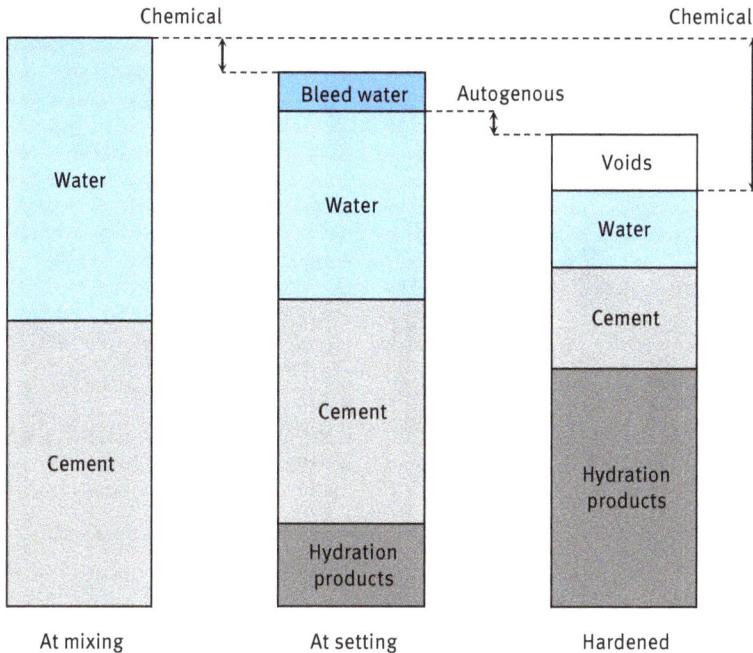

Figure 5.2: Definition of chemical and autogenous shrinkage.

autogenous shrinkage are theoretically shown in Figure 5.2 and autogenous shrinkage will be the focus point in this section.

The formed hydrostatic tensile forces, especially in systems with a low water-to-binder ratio, induce cracking. At first, these small and narrow cracks do not seem to impose such a big problem but intruding substances may cause failure of the material. The shrinkage is caused by the lowering of the internal RH [1, 2, 33] and self-desiccation when no external water source is present. The internal microcracks may interconnect flow paths for penetrating water and gases, possibly containing harmful substances during the service life of concrete structures. By maintaining the internal RH, this can be counteracted. That is why SAPs were first used to mitigate autogenous shrinkage due to their internal curing effect. The application of SAPs for this purpose proved to be successful as autogenous shrinkage was reduced and even counteracted in time [1, 2, 21–25].

The principle of the SAPs for internal curing is found in Figure 5.3. During preparation of a cementitious mixture, the SAPs will take up mixing water. The SAPs will form water-filled inclusions, useful for internal curing [21] as the water is released again in time. The water released due to self-desiccation during cement hydration can be used for further hydration and reduction of the autogenous shrinkage [22]. The water present in the SAP will hereby be released into the cementitious matrix due to the imminent drop in RH. Due to this water release, the internal

Fresh state   Hardened

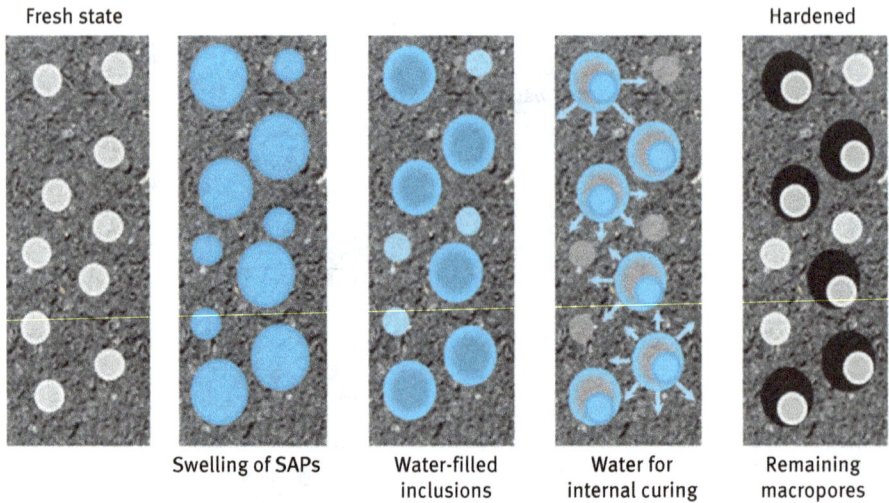

Swelling of SAPs   Water-filled   Water for   Remaining
                        inclusions   internal curing   macropores

Figure 5.3: Different steps for internal curing when SAPs are used in cementitious materials [8].

RH is maintained. The SAP particles shrink and an empty macropore remains as shown by means of neutron tomography measurements [34–38]. The macropore showed a densification around its perimeter [39–41]. Due to the internal curing the autogenous shrinkage can be completely mitigated in systems with pure cement, combined with silica fume, fly ash, and blast furnace slag [10, 23, 24, 42–44].

For internal curing, the water kinetics of the SAPs are important [45]. If this water is released too soon, it leads to a significant decrease in compressive strength. But if this water is released at the ideal stage (beginning of concrete setting as the earliest point), this water would serve as internal curing water [18]. It is very important to use a SAP with the ideal properties. If the water is released too fast (i.e., before setting), the microstructure will be completely different and if the water is released too late (i.e., after a couple of days onwards), the purpose of internal curing vanishes. This was studied in detail using NMR where the entrained water signal was distinguishable from the free water in the cementitious system. In time, the water released from the SAPs toward the cementitious matrix could be studied [45]. More recently, elastic wave nondestructive testing may also be a way to monitor the water kinetics by the SAPs [46].

Typical amounts of 0.2–0.6 m% of cement weight of SAPs are used [3]. The amounts are based on the theory of powers [47] stating the amount of additional water needed to counteract autogenous shrinkage. Again, the type of polymer is important, as the absorption and release kinetics in a cementitious environment are different [33, 48, 49]. The type of polymer and the interaction with specific ions and cations play a role in terms of the absorption and release kinetics and were less related to the cross-linking density [18].

5.4 Changing the microstructure to increase the freeze–thaw resistance by the formation of an internal void system

As soon as cement and water come into contact, hydration reactions start. The hydration of a concrete mixture determines the microstructural development and SAPs influence this formation. This was already extensively studied. Hardened mixtures with SAPs showed less capillary porosity at later ages if additional water was used (compared to if no additional water was used). The water released from the SAPs resulted in continued hydration, decreasing the microporosity at later ages [23, 50], except from the macropores created by the SAPs. A reduction of the amount of smaller capillary pores was seen [51]. This is due to two effects: (1) the filling of the existing pores with hydration products due to internal curing [26] and (2) the reduction of the initial microcracks in the interior of a cementitious matrix, as autogenous shrinkage is partially reduced. Mixtures with the same effective water-to-cement ratio (ratio of the mixing water not held by the SAPs over the cement content) showed the same capillary porosity [52–55]. The microstructure in between SAPs was denser due to internal curing and the possible stimulated additional hydration caused by this release of water. The structure of a cementitious material was affected by the apparent water-to-cement ratio. As SAPs take up the mixing water, the apparent water-to-cement ratio appears lower, resulting in a closely packed matrix and subsequent hydration due to the release of the stored water. Samples without SAPs do not have access to this stored water. Therefore, the permeability was lower in between SAP macropores in samples containing SAPs than of reference samples. This was also shown by using neutron radiography [35, 36] and supported by modeling on mesoscale level [56, 57].

As can be expected due to the swollen size of the SAPs and the remaining macroporosity, mixtures with SAPs showed a higher total porosity due to macropore formation when additional water was used [21, 39, 58]. If no additional water was added, the total porosity may be lower for mixtures with SAPs [39] as the overall porosity decreased due to the densification even though macropores remain.

Microstructural properties, and especially the macropore formation, directly affect the strength characteristics of the cementitious material. The flexural and compressive strength decrease when SAPs and additional water are added [2, 3, 21–23, 36, 48, 53, 59–63]. Internal curing leads to further hydration and the effect of SAPs on strength loss is reduced at later ages. The further hydration improves the mechanical properties but is mostly counteracted by the strength loss caused by macropore formation due to the absorption of mixing water by the SAPs [64]. SAPs thus have both a positive and a negative effect on the mechanical properties. A decrease in strength is observed at earlier testing ages (<7 days) while sometimes increases are obtained at later ages [65], especially in systems with supplementary cementitious materials where the internal curing reservoirs are available for the longer term pozzolanic reactions. These characteristics depend on the polymer, mixture composition, water-to-binder ratio, amount of additional water, concrete versus mortar or paste, amount of SAPs added, curing conditions, testing age, and so on. For example, an amount of 0.2–0.6 m% versus cement weight of SAP was used to reduce the autogenous shrinkage [3], while for sealing and healing purposes, this amount was up to 1 m% [35, 66, 67]. This will influence the impact on the observed mechanical properties. Generally, in literature, a decrease in compressive strength is found [18, 26, 48, 68–73] as there is a change in microstructure [53, 55, 58, 68, 71, 72, 74–76]. When no additional water is added, there is a shift in effective water-to-cement ratio and a possible densification of the cementitious matrix. One should be very careful when comparing the mechanical properties of these different cementitious systems. Typical values are a decrease of 10–20% for acrylate SAPs with a size of 300–500 μm and 30–50% for smaller SAPs with a size of 50–150 μm [53, 68] when 0.2–0.5 m% of SAPs are added. Even though the system of macropores reduces the mechanical properties, this property is interesting considering the improvement of the impact strength in strain-hardening mixtures [60]. The macropores serve as stress activators, increasing the ductility [77] and impact absorbing features [60].

To limit the influence of the swelling SAPs on the mechanical properties, pH-sensitive SAPs [78–82] or coated SAPs [83, 84] may be used. Alginates, for example, do not reduce the strength due to their low absorption capacity [82, 85]. This lower swelling capacity is interesting in order to limit the absorption in the initial stage and aiming at other applications such as sealing and healing, needing a later swelling capacity at later ages [66, 79, 80, 82, 83]. The strength can also be compensated by the use of colloidal silica nanoparticles upon addition of SAPs [86–88]. The strength-loss due to the macropore formation is compensated by incorporating these nanomaterials which strengthen the overall cementitious matrix.

As SAPs create an internal void system (Figure 5.4), they increase the freeze–thaw resistance if properly designed [39, 40, 58, 89–94]. The voids act in the same way as if an air-entraining agent is added. In case of this internal void system, the freezing water has a pathway to expand, limiting the formation of cracks, scaling, and general expansion of the cementitious matrix. Compared to a system with an air-entraining agent, the SAP mixtures increase the freeze–thaw resistance without extreme strength loss [89] and with proper mix stability. When using an air-entraining agent, the air bubbles may migrate upon long mixing times. The SAPs are thus an interesting material to add to the cementitious matrix. As the absorption capacity in the cementitious matrix is known, the formed macroporosity can be designed to have the optimal sizes and spacing factors for a specific application. The addition of SAPs in the range of 0.10–0.34% in relation to the mass of cement has been reported to promote a reduction of at least 50% in the scaling after more than 25 freeze–thaw cycles in both cement mortars and concrete mixtures [95, 96]. Not only the amount of SAPs but also their particle size and production process might have

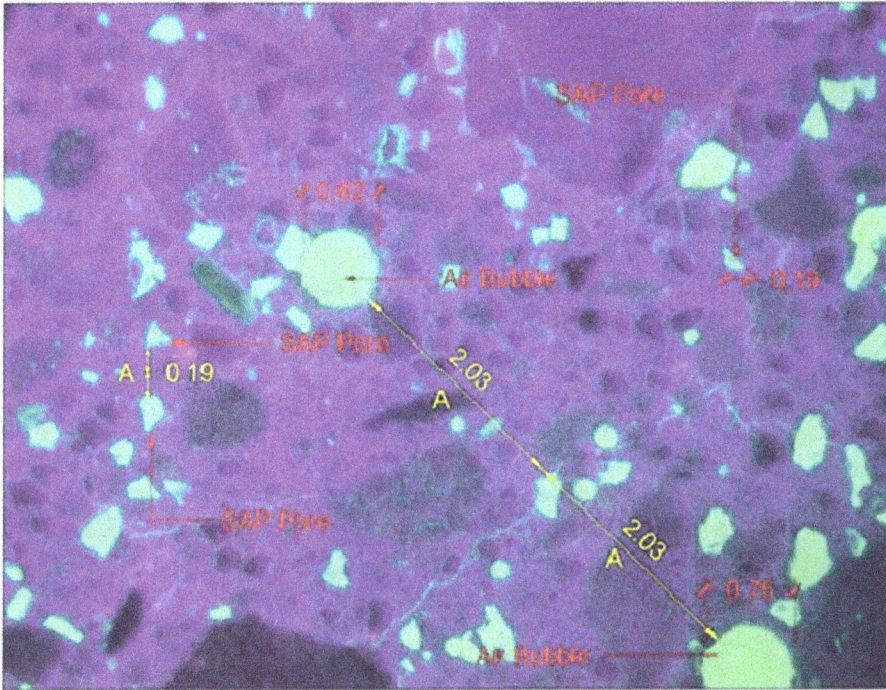

Figure 5.4: Internal void system with air bubbles and SAP pores, for increasing the freeze–thaw resistance, using a thin section and fluorescent light microscopy. This picture has been partly redrafted with (CC BY) license from [94].

an impact on the scaling resistance. In addition, the time of adding the polymers during plant-scale mixing is of importance. The addition of SAPs directly in the truck, after the mixing procedure at the plant mixer, showed no significant impact on the compressive strength of the concrete but an agglomeration of air void particles and an inferior performance in terms of shrinkage reduction occurred [33, 49, 97]. Adding SAPs on the materials' belt, along with the dry materials, or in water-soluble bags has shown promising results. This is of importance when using a specific concrete for road construction applications.

5.5 Regaining the water impermeability through self-sealing of cracks

Due to their swelling capacity upon contact with fluids, SAPs may cause a decrease in permeability of cracked cementitious materials. When liquids enter a crack, SAP particles along the crack faces will swell and block the crack [7, 8, 35, 36, 38, 66, 98–105].

In this way, the impermeability of cracked cementitious materials can be regained (Figure 5.5). Application of a superabsorbent resin in situ to repair concrete leakage can also be used, but this is rather considered to be manual applications [7, 106] while mixed in SAPs are always present to immediately seal the occurring cracks. In 100–300 μm wide cracks, SAPs with a size of approximately 500 μm were better in terms of sealing compared to 100-μm-sized SAPs as the latter were washed out and were not able to fill cracks, even though high amounts (1 m% of cement weight) were used [8, 35, 67]. It was also found that due to the swelling effect of the SAPs, the reduced water movement speed, which was critical to obtain autogenous healing, was optimal as cracks were able to close due to deposited crystals. In reference specimens, the amount of autogenous healing – inherent part of a cementitious system, see later on – was less compared to the specimens with SAPs. In water-retaining structures like quays or cellars the SAPs may prove to be useful as the flow will be reduced, sealing the cracks, but the crack may be sealed by deposited crystals as well. This is also the case in large-scale specimens or observed underneath bridges and in tunnels [107]. As studied by cryofracture scanning electron microscopy, SAPs swell across voids including cracks, causing a sealing of the cementitious material [103].

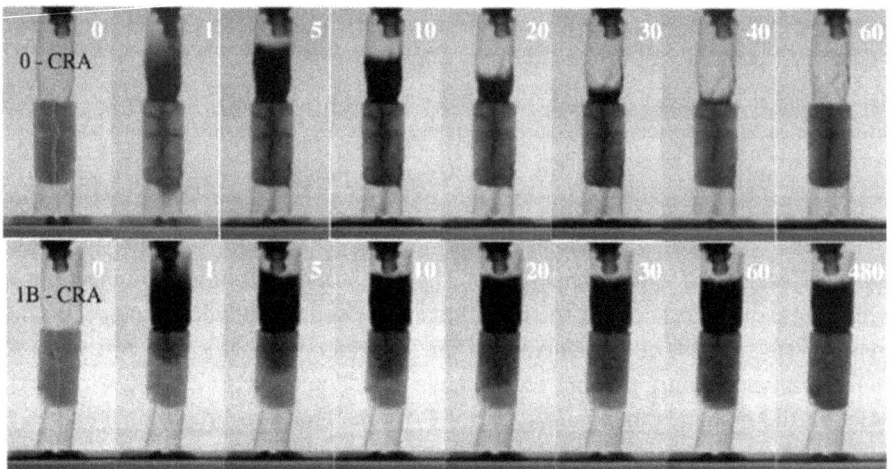

Figure 5.5: Self-sealing concrete showing a cracked cementitious material with a 1 cm diameter and 20 mm height without SAPs (**top**) and with SAPs (**bottom**). The imposed water head is not stopped in the reference material while a sealing effect is noticed with SAPs due to their swelling ability. The time is mentioned in the upper right corner in seconds. This picture has been partly redrafted with permission from Elsevier from source [35].

Self-sealing is related to the initial decrease in permeability and is not permanent. This is important to know, as a possible temporal self-sealing effect may not lead to a regain in mechanical properties. This regain, on the other hand, is the result of self-healing.

5.6 Regaining the mechanical properties due to promoted autogenous healing

Concrete already possesses the natural capacity of autogenous crack healing [4], as first found by the French Academy of Science in 1836 (as stated in [108]). It seems strange that this solid and gray material possesses this feature, but it can be seen everywhere around us. When passing underneath a concrete bridge or through a concrete tunnel, whitish crystals can be seen near and on cracks throughout the material. This is considered to be healing. One can design a concrete material to include an additional healing capacity, the so-called autonomous healing (such as polymeric foams and vascular systems) but autogenous healing is also inherent to concrete. This latter term means that concrete is able to heal its own crack, using its initial constituents or already formed products.

Four main mechanisms and their combined effect contribute to autogenous healing of concrete cracks [108–117] (as shown in Figure 5.6):

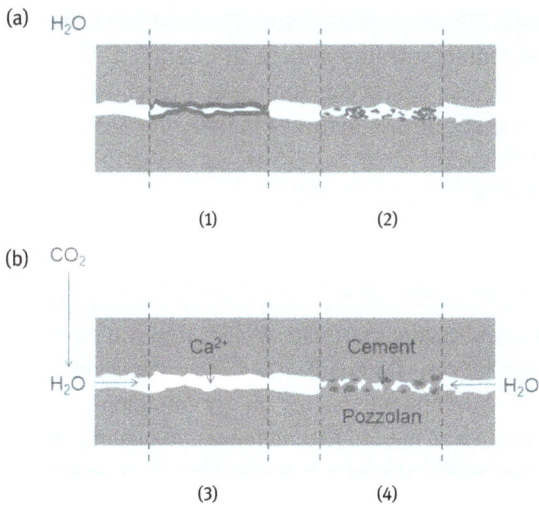

Figure 5.6: The four different healing mechanisms responsible for autogenous crack healing: **(1)** expansion of calcium silicate hydrates, **(2)** blockage by loose particles, **(3)** crystallization of calcium carbonate, and **(4)** further hydration combined with pozzolanic activity [8].

1) The matrix may expand due to swelling of calcium silicate hydrates (C-S-H) as the layering system of the gel becomes wider.
2) Loose and broken-off particles or impurities in the matrix, fluid, or surroundings may block the crack.

3) Dissolved carbon dioxide from the ambient air in water may react with Ca^{2+} ions present from hydration products in the concrete matrix to form the often observed white calcium carbonate ($CaCO_3$) crystals.
4) Unhydrated cement grains present in the matrix and on the crack surfaces may further hydrate when these particles are exposed to water. In addition, supplementary cementitious materials such fly ash or blast furnace slag can still react through pozzolanic or latent hydraulic activity. Pozzolans promote further hydration as these materials react with water and $Ca(OH)_2$ to form C-S-H.

The first two mechanisms (C-S-H expansion and blockage by impurities and other particles) are the inferior ones while the further hydration and calcium carbonate crystallization are the dominant mechanisms in order to receive a strength regain. The strength is mainly gained by further hydration as $CaCO_3$ crystals do not have sufficient strength compared to the cementitious hydration material [118]. However, the white crystals are most often observed, in combination with the grayish further hydration [111, 112, 119, 120], providing an aid in sustained promoted autogenous healing, even up to several years [118]. In high-strength concretes showing a low water-to-cement ratio, the healing is mainly due to the hydration of unhydrated cement grains on the crack surfaces as more unhydrated cement remains present [113, 121, 122]. Also, the younger the material is, the more healing will occur due to the higher amount of unhydrated particles [118]. As the cement further hydrates in time, the healing material formed at early ages is a combination of $CaCO_3$, C-S-H, and $Ca(OH)_2$. At later ages, the healing material is mainly $CaCO_3$ [110, 114, 118].

Assuming that specific chemical substances (Ca^{2+}, CO_2, etc.) are present in the mixture composition or from the specific hydration products, the exposure to humid environmental conditions (wet/dry cycles, submersion in water, etc.) and restricted crack widths up to 30–50 µm for strain-hardening mixtures [67, 109] are the main areas of focus. Only when building blocks, abundant water, and restricted crack widths are present, the material may show optimal healing. In dry conditions, that is, without the presence of liquid water such as at 95% RH, there was no healing visible and it was concluded that the presence of water as a curing medium was essential. As water is needed in all mechanisms [109, 111, 113, 123, 124], the role of SAPs becomes clear [67, 125]. Of course, the crack width and mixture composition play a huge role. Pozzolanic fly ash [126, 127], blast-furnace slag [128], lime [129], or alkaline activators [130] can be added to receive more autogenous healing. Additives like expandable geomaterials [131] or crystalline admixtures [132–136] stimulate the crack-healing capacity even further. The use of SAPs has also been explored in combination with expansive agents such as calcium sulfoaluminate in sulfur composites [137]. The stimulated autogenous healing has also been studied in specimens containing pH-sensitive SAPs or natural polymers in combination with a synthetic backbone [79, 80, 82, 85, 138]. In order to further increase the amount of calcium carbonate precipitation in a wide crack of several hundreds of micrometers, SAPs can be combined with bacteria [139, 140]. The cross-linking of the SAPs is performed after addition of carbonate-precipitating bacterium, such as *Bacillus sphaericus*, in order to properly entrap them and protect them from the harsh alkaline environment.

As the cementitious material has a problem with healing large "fractures" or "cuts" like the human body, the crack width should be restricted. This can be achieved by adding synthetic microfibers to the cementitious mixture [4, 67, 109, 123, 141–148], or by using natural fibers as a greener solution [149, 150]. The use of glass fibers can even be combined to have additional translucent properties of the gray cementitious material [151]. So, as mentioned, SAPs can be added to further stimulate the autogenous healing [8, 38, 60, 61, 66, 67, 107, 118, 125, 152–155], due to their retentive capacity as shown in Figure 5.7. Closely resembling bone healing, the links are made toward cementitious healing of narrow cracks. When a crack occurs, the SAPs will be exposed to the environment. They will start to absorb a fluid upon contact and/or the SAPs will start to adsorb moisture from the environment. This will cause a physical sealing of the crack, slowing down the fluid flow through a crack. This is related to the bleeding and clogging found in the human body. The SAPs will release their absorbed water for stimulating the autogenous healing mechanisms, especially during the drier periods. This process, like human bone reconstruction, continues until the complete crack is closed or the building blocks are consumed or exhausted. In the end, and the ideal case, a healed cementitious material is obtained with the same or even better mechanical properties compared to an uncracked material. SAPs will remain present at their location and will be available for subsequent healing if the conditions are again optimal with sufficient building blocks, water or moisture, and narrow healable cracks.

When not completely submerged in water, only samples containing SAPs showed self-healing properties due to moisture uptake [36, 67]. Even in an environment with RH > 90%, there was noticeable healing, due to their moisture uptake capacity. If reference samples were stored in a climate room with a certain RH, there was almost no autogenous healing, as water was not present to form the healing products. The samples with SAPs showed a regain in strength when stored in an RH of more than 90%. The moisture uptake by SAPs (up to four times their own weight in moisture [55]) seemed to be sufficient to promote a certain degree of autogenous healing, especially in the interior of the crack in the form of further hydration. In the RH condition of more than 90%, the material with 1 m% showed a regain of 60%. At the crack mouth, the crack was still clearly open and only at some distinct places, there was some bridging of a crack by healing products.

Cracks smaller than 30 µm exposed to wet/dry cycles healed completely both with and without SAPs after a healing period of 28 days. SAPs can contribute to the internal healing of a crack after performing wet/dry cycles [67, 125]. Cracks between 50 and 150 µm healed partly in samples without SAP, but sometimes even some cracks closed completely after 28 wet/dry cycles in a specimen containing SAPs [67], as shown in Figure 5.8. Cracks larger than 200 µm showed almost no healing.

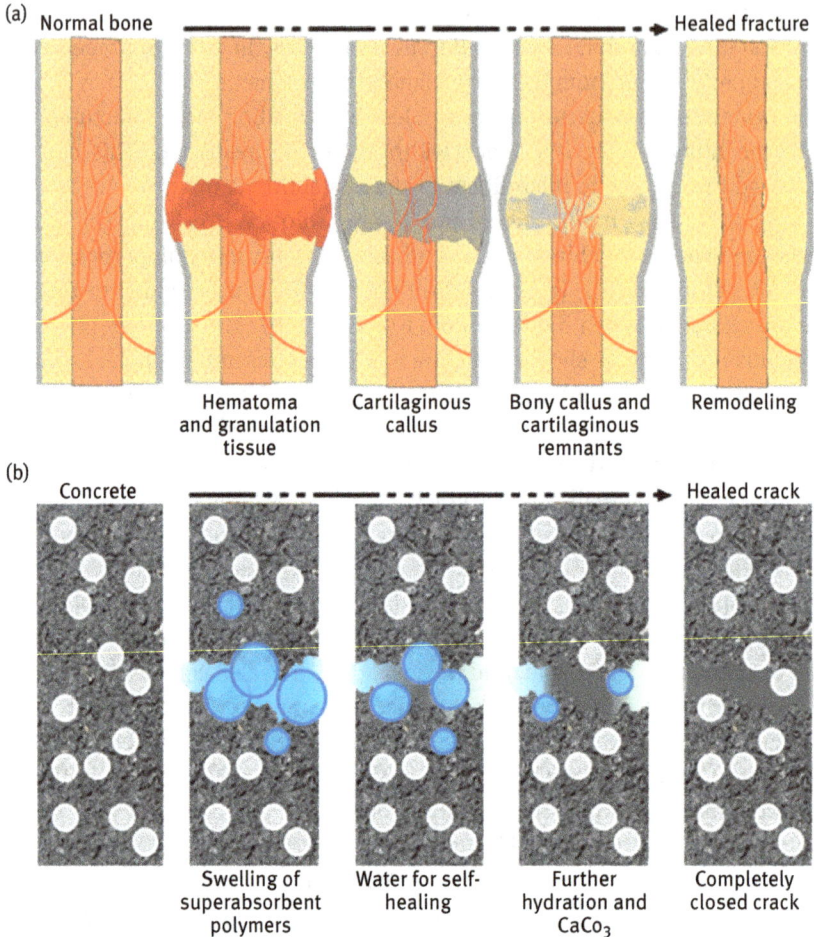

Figure 5.7: Use of superabsorbent polymers in cementitious materials to stimulate autogenous healing, as a biomimicry of bone healing found in living creatures [8].

The cracks are considered too wide to be healed properly within a 28-day period, even though SAPs are promoting autogenous healing.

This healing product formation could be visualized by means of X-ray microtomography [155], as shown in Figure 5.9. The figure shows horizontal slices of small 6 mm wide samples without (top) and with SAPs (bottom), stored at high RH (>90%, left) and in wet/dry cycling (right). The cracks are clearly seen in black, together with the spherical air porosity. The irregular-shaped voids in the bottom part of the figure are macropores formed by the swollen SAP right after final setting. The yellow colors in the figure show the material that is adhered after performing the 28-day healing cycle. This information was obtained by comparing the microtomograms prior and

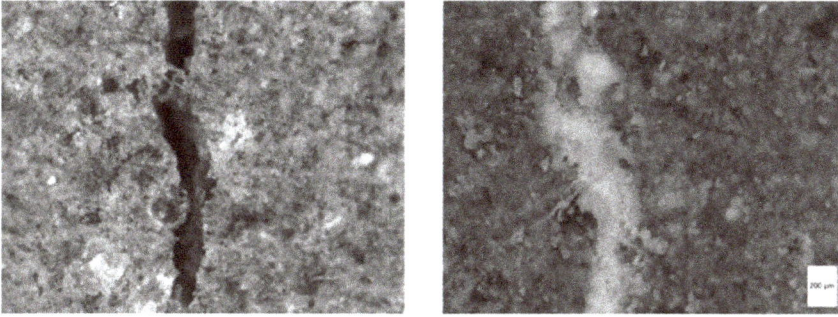

Figure 5.8: Example of a specimen containing 1 m% of SAP after performing 28 days of wet/dry cycling. The whitish healing products are mainly calcium carbonate with some further hydration [8].

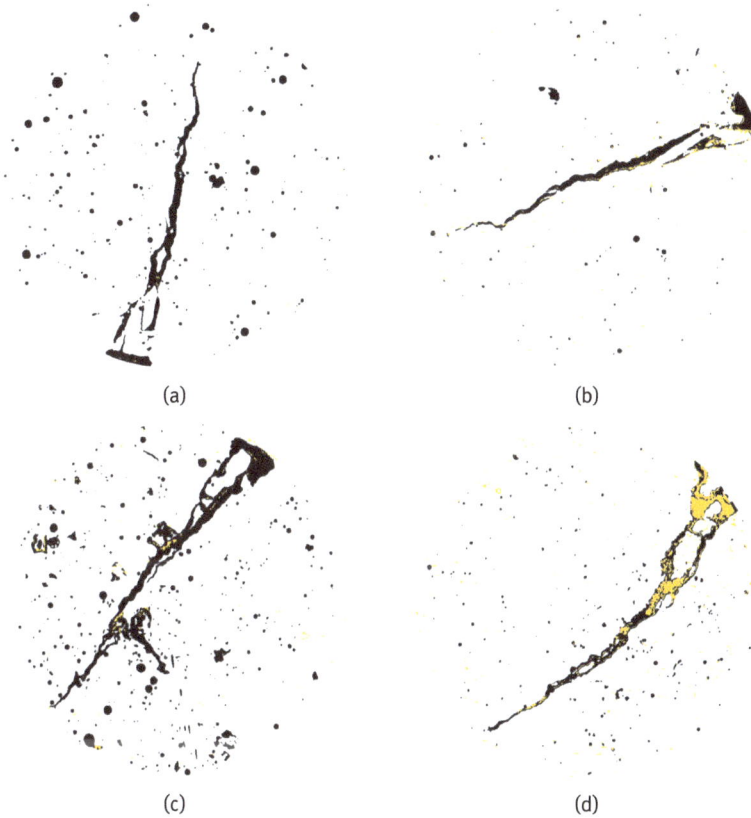

(a)

(b)

(c)

(d)

Figure 5.9: (a, b) Cross sections of the specimens without (c, d) and with (a, c) superabsorbent polymers stored at more than 90% RH (b, d) and in wet/dry cycling. Black depicts the porosity and the crack, and yellow visualizes the formed healing products. The diameter of the specimen is 6 mm. This picture has been partly redrafted with permission from Elsevier from source [155].

after healing. No healing was observed in specimens without SAPs when no liquid water is present (Figure 5.9a), while some crystal formation was present near the vicinity of SAPs in SAP specimens (Figure 5.9c). This amount of healing was comparable to the amounts found in SAP-less specimens healed in wet–dry cycling (Figure 5.9b). The largest amount of healing was observed in specimens with SAPs and stored in wet/dry cycling (Figure 5.9d). Almost the complete crack was closed. Some other conclusions could also be drawn in the research performed. The largest amount of healing was found in the region 0–100 μm below the surface. In the interior of a crack, the amount of healing products was less and only at some distinct places, the healing products bridge a crack, probably in the vicinity of a fiber (as they act as a nucleation site for the calcium carbonate crystals [8, 118]). The healing at high RH occurred in the vicinity of the SAP particles, stitching the crack at distinct locations [155]. This healing, as well as in wet/dry cycling, was still stimulated in samples with 8 years of age [118].

The autogenous healing capability of cementitious materials was maintained during subsequent loading cycles to a certain degree. SAP particles promote the self-healing ability by renewed internal water storage upon crack formation and this leads to regain mechanical properties such as the first cracking strength. In wet/dry cycles, the plain material without SAPs was able to regain 45% of its first cracking strength after a first healing cycle. After the second healing cycle, this regain was 28%. When SAPs were used, the regain was 75% and 66%, respectively [61]. The better healing in specimens containing SAPs during first and second loading and healing was also confirmed by natural frequency analysis [60]. Possible explanations are the storage of a calcium-rich fluid (i.e., the pore solution) in the swollen SAPs, and the reduced permeation through the crack. This provides the possibility of the formation of the $CaCO_3$ crystals in the crack. This caused the ideal conditions for promoted autogenous healing. The regain in mechanical properties was noteworthy.

The promoted healing capacity in systems with SAPs was also studied by means of NMR testing [156]. This will prove to be essential for model verification and to increase the simulations for this new type of material in future research. Adding 1 m% of SAP to a cementitious material stimulated further hydration with nearly 40% in comparison with a cementitious material without SAPs, in wet–dry cycling. At 90% RH, no healing was observed in reference samples while specimens with SAPs showed up to 68% of further hydration compared to a reference system without SAPs healed in wet/dry cycling, due to the uptake of moisture by the SAPs. This proved the differences in observed regain in mechanical properties.

5.7 Practical applications of SAPs in the construction industry

In 2006, a pavilion was built for the FIFA World Cup [3, 157]. It was a thin-walled structure with very slender columns without conventional reinforcement. The low water-to-binder structure did not show any kind of cracks due to the prevention of autogenous shrinkage cracks by the included SAPs and the sustained internal curing.

In China, SAPs have already been applied in several civil engineering construction projects. Examples are the Lanzhou–Urumqi railway where SAPs were used to prevent massive concrete slabs from cracking. The concretes with SAPs were less sensitive to moisture loss because of evaporation [63, 158]. Two other field applications were performed in southern China [159]. A shear wall structure that had dimensions of 20 × 50 x 0.85 m cast in one time did not show surface cracks after 7 days when SAPs were added. A cast-in-place concrete floor slab was cast in one time and had dimensions of 12 × 8 x 0.12 m^3 and no curing methods were used. No cracks were observed and the used gel-type SAP showed potential to mitigate early-age cracking.. SAPs were also applied in the China Zun tower [160]. It was found that the SAPs reduced shrinkage by 46% and that the later age compressive strength was not affected when 0.56 m% of SAP was added to the mixture.

A project on large-scale tunnel elements in Belgium, iSAP, will include the application of large-scale elements for tunnel construction [97]. SAPs will hereby be added as a mitigation measure for autogenous shrinkage. In this way, occurring shrinkage cracks may be avoided. The SAPs will also possess self-sealing and self-healing properties. In this way, some remaining shrinkage cracks may be sealed and healed and the structure will possess some sealing potential for observed cracking during its lifetime. This is interesting for the application in tunnels and other ground-retaining structures, as the flow of harmful substances is stopped [33, 49].

5.8 Conclusions

In conclusion, there is still a lot of unraveling to do in order to apply SAPs into practice. Possible applications of the self-sealing and self-healing material are widely spread. Water-retaining structures may benefit and construction companies may be interested. The principle of using SAPs has its possible applications for the industry. Contractors are searching for a way to decrease shrinkage cracks and to obtain a watertight structure. This is especially important for tunnel elements, underground parking garages, basements, liquid containing structures, pavements, and so on. Nowadays, contractors are often forced to apply crack repair right after construction, due to the formation of shrinkage cracks and thermal cracks at early age. The shrinkage could be overcome by using SAPs as they may provide internal curing to the construction element: they absorb water in the fresh concrete mix, and provide it to the cement particles at the right moment in the hydration process, in this way reducing the autogenous shrinkage. In hardened concrete, they may seal occurring cracks, as they swell in contact with intruding water. This may reduce the uptake of harmful substances, most likely leading to an enhanced long-term durability and service life. The SAPs will subsequently promote autogenous healing of the crack since they provide water for further hydration of yet unhydrated cement particles and calcium carbonate precipitation, leading to even more tight structures and possible regain of the mechanical properties. More research is needed in terms of the long-term durability of these novel cementitious materials with SAPs [161].

Microfiber-reinforced strain-hardening cementitious composites possess the qualities of a high-strength concrete combined with tensile ductility and crack width control. Their small cracks are interesting in terms of autogenous healing where only small cracks are able to heal completely, further stimulated by SAPs. Combined with (promoted) self-healing it is a durable material and very promising to use in the future. In regions with wet/dry cycles, water remains present in the SAPs during the dry periods. Therefore, self-healing can prevail at all times. However, performance-based durability concepts are still required to get a durability design framework for these strain-hardening materials [162].

Furthermore, due to the self-compacting properties of the strain-hardening mixture, thin forms are achievable [120]. Nature fits form to function. This is also true for this material; the accretion of material to places where it is most needed, resulting in adaptive structures. The form should be ideal to transfer loads, so that an excess of material can be removed. This material will result in lighter and safer structures, leading to a reduced safety factor as the structure may reach its optimal design.

The role of autogenous healing on corrosion prevention will also be important in the future. If cracks are not sealed, water containing aggressive substances will break down the passive film on the reinforcements. This aspect needs to be considered when autogenous healing is used in real-life structures. The maintenance and longevity of these structures is hereby very important. The close investigation on plastic shrinkage mitigation and other promising pathways for inclusion of SAPs in cementitious materials are also key for the near future. This white powder thus will be more accepted in the conservative building industry.

One general conclusion can be made; one should continue to build with nature's rules. The bleeding (water for SAPs), blood cells (building blocks), blood flow vascular network (porous concrete), blood clothing (formation of healing products near synthetic microfibers), skeleton and bone healing (crystallization) are only a few properties studied in the field of construction healing. By mimicking nature to enhance performance, constructions that are more durable will be designed, leading to a higher service life and better overall life quality.

References

[1] Jensen O.M., Hansen P.F. Water-entrained cement-based materials I. Principles and theoretical background. Cem Concr Res 2001, 31, 647–654.

[2] Jensen O.M., Hansen P.F. Water-entrained cement-based materials II. Experimental observations. Cem Concr Res 2002, 32, 973–978.

[3] Mechtcherine V., Reinhardt H.W. Application of Super Absorbent Polymers (SAP) in concrete construction. In: State-of-the-Art Report Prepared by Technical Committee 225-SAP. Springer, 2012, 165.

[4] Snoeck D., De Belie N. From straw in bricks to modern use of microfibres in cementitious composites for improved autogenous healing – a review. Constr Build Mater 2015, 95, 774–787.

[5] Schröfl C., Snoeck D., Mechtcherine V. A review of characterisation methods for superabsorbent polymer (SAP) samples to be used in cement-based construction materials – Report of the RILEM TC 260-RSC. Mater Struct 2017, 50, 1–19.

[6] Mignon A., Snoeck D., Dubruel P., Van Vlierberghe S., De Belie N. Crack mitigation in concrete: superabsorbent polymers as key to success? Mater 2017, 10, 1–25.

[7] Snoeck D., Malm F., Cnudde V., Grosse C.U., Van Tittelboom K. Validation of self-healing properties of construction materials through non-destructive and minimal invasive testing. Adv Mater Interfaces 2018, 1800179.

[8] Snoeck D. Self-healing and microstructure of cementitious materials with microfibres and superabsorbent polymers. In: Struct Eng. Ghent, Ghent University, 2015, 364.

[9] Snoeck D., Schröfl C., Mechtcherine V. Recommendation of RILEM TC 260-RSC: testing sorption by superabsorbent polymers (SAP) prior to implementation in cement-based material. Mater Struct 2018, 51, 1–7.

[10] Wyrzykowski M., Igarashi S.-I., Lura P., Mechtcherine V. Recommendation of RILEM TC 260-RSC: using superabsorbent polymers (SAP) to mitigate autogenous shrinkage. Mater Struct 2018, 51, 1–7.

[11] Mechtcherine V., Schröfl C., Reichardt M., Klemm A.J., Khayat K.H. Recommendations of RILEM TC 260-RSC for using superabsorbent polymers (SAP) for improving freeze-thaw resistance of cement-based materials. Mater Struct 2019, 52, 1–7.

[12] Mechtcherine V., Snoeck D., Schröfl C., De Belie N., Klemm A.J., Ichimiya K., Moon J., Wyrzykowski M., Lura P., Toropovs N., Assmann A., Igarashi S., De La Varga I., Almeida F.C.R., Erk K.A., Ribeiro A.B., Custódio J., Reinhardt H.W., Falikman V. Testing superabsorbent polymer (SAP) sorption properties prior to implementation in concrete: results of a RILEM Round-Robin Test. Mater Struct 2018.

[13] Zohuriaan-Mehr M.J., Kabiri K. Superabsorbent Polymer Materials: a Review. Iran Polym J 2008, 17, 451–477.

[14] Vandenhaute M., Snoeck D., Vanderleyden E., De Belie N., Van Vlierberghe S., Dubruel P. Stability of Pluronic® F127 bismethacrylate hydrogels: reality or utopia? Polym Degrad Stab 2017, 146, 201–211.

[15] Jensen O.M. Use of superabsorbent polymers in construction materials. In: Sun W., van Breugel K., Miao C., Ye G., Chen H., Eds. International Conference on Microstructure Related Durability of Cementitious Composites. Nanjing, RILEM Publications S.A.R.L., 2008, 757–764.

[16] Mechtcherine V., Secrieru E., Schröfl C. Effect of superabsorbent polymers (SAPs) on rheological properties of fresh cement-based mortars – Development of yield stress and plastic viscosity over time. Cem Concr Res 2015, 67, 52–65.

[17] Secrieru E., Mechtcherine V., Schröfl C., Borin D. Rheological characterisation and prediction of pumpability of strain-hardening cement-based composites (SHCC) with and without addition of superabsorbent polymers (SAP) at various temperatures. Constr Build Mater 2016, 112, 581–594.

[18] Schröfl C., Mechtcherine V., Gorges M. Relation between the molecular structure and the efficiency of superabsorbent polymers (SAP) as concrete admixture to mitigate autogenous shrinkage. Cem Concr Res 2012, 42, 865–873.

[19] Van Der Putten J., Azima M., Van den Heede P., Van Mullem T., Snoeck D., Carminati C., Hovind J., Trtik P., De Schutter G., Van Tittelboom K. Neutron radiography to study the water ingress via the interlayer of 3D printed cementitious materials for continuous layering. Constr Build Mater 2020, 258, 119587.

[20] Schröfl C., Nerella V.N., Mechtcherine V. Capillary water intake by 3D-printed concrete visualised and quantified by neutron radiography. In: Wangler T., Flatt R., Eds. First RILEM International Conference on Concrete and Digital Fabrication – Digital Concrete. Cham, Zürich, Springer, 2019, 217–224.

[21] Mechtcherine V., Dudziak L., Hempel S. Mitigating early age shrinkage of ultra-high performance Concrete by using Super Absorbent Polymers (SAP). In: Sato R., Maekawa K., Tanabe T., Sakata K., Nakamura H., Mihashi H., Eds. Creep, Shrinkage and Durability Mechanics of Concrete and Concrete Structures. Ise-Shima, Taylor & Francis, 2009, 847–853.

[22] Craeye B., De Schutter G. Experimental Evaluation of Mitigation of Autogenous Shrinkage By Means of a Vertical Dilatometer for Concrete. In: Jensen O.M., Lura P., Kovler K., Eds. International RILEM conference on Volume changes of hardening concrete: testing and mitigation. Lyngby, RILEM Publications S.A.R.L., 2006, 21–30.

[23] Igarashi S., Watanabe A. Experimental study on prevention of autogenous deformation by internal curing using super-absorbent polymer particles. In: Jensen O.M., Lura P., Kovler K., Eds. International RILEM Conference on Volume Changes of Hardening Concrete: testing and Mitigation. Lyngby, RILEM Publications S.A.R.L., 2006, 77–86.

[24] Mechtcherine V., Gorges M., Schröfl C., Assmann A., Brameshuber W., Bettencourt Ribeiro V., Cusson D., Custódio J., Fonseca da Silva E., Ichimiya K., Igarashi S., Klemm A., Kovler K., Lopes A., Lura P., Nguyen V.T., Reinhardt H.W.T.F., Weiss R.D.J., Wyrzykowski M., Ye G., Zhutovsky S. Effect of internal curing by using Superabsorbent Polymers (SAP) on autogenous shrinkage and other properties of a high-performance fine-grained concrete: results of a Rilem Round-Robin Test, TC 225-SAP. Mater Struct 2014, 47, 541–562.

[25] Assmann A. Physical properties of concrete modified with superabsorbent polymers. In: Civil and Environmental Engineering. Stuttgart, Stuttgart University, 2013, 199.

[26] Justs J., Wyrzykowski M., Bajare D., Lura P. Internal curing by superabsorbent polymers in ultra-high performance concrete. Cem Concr Res 2015, 76, 82–90.

[27] Dudziak L., Mechtcherine V., Enhancing early-age resistance to cracking in high-strength cement based materials by means of internal curing using super absorbent polymers, in: Brameshuber W. (Ed.) International RILEM Conference on Material Science, RILEM Publications S.A.R.L., Aachen, 2010, pp. 129–139.

[28] Soroka I. Concrete in Hot Environments. London, E & FN Spon, 1993.

[29] Combrinck R., Boshoff W.P. Typical plastic shrinkage cracking behaviour of concrete. Mag Concr Res 2013, 65, 486–493.

[30] Snoeck D., Pel L., De Belie N. Superabsorbent polymers to mitigate plastic drying shrinkage in a cement paste as studied by NMR. Cem Concr Comp 2018, 93, 54–62.

[31] Boshoff W.P., Mechtcherine V., Snoeck D., Schröfl C., De Belie N., Ribeiro A.B., Cusson D., Wyrzykowski M., Toropovs N., Lura P. The effect of superabsorbent polymers on the mitigation of plastic shrinkage cracking of conventional concrete, results of an inter-laboratory test by RILEM TC 260-RSC. Mater Struct 2020, 53, 1–16.

[32] Schröfl C., Mechtcherine V., Vontobel P., Hovind J., Lehmann E. Sorption kinetics of superabsorbent polymers (SAPs) in fresh Portland cement-based pastes visualized and

quantified by neutron radiography and correlated to the progress of cement hydration. Cem Concr Res 2015, 75, 1–13.

[33] Tenório Filho J.R., Araújo M.A., Snoeck D., De Belie N. Discussing different approaches for the time-zero as start for autogenous shrinkage in cement pastes containing superabsorbent polymers. Mater 2019, 12, 2962.

[34] Trtik P., Muench B., Weiss W.J., Herth G., Kaestner A., Lehmann E., Lura P. Neutron tomography measurements of water release from superabsorbent polymers in cement paste. In: Brameshuber W., Ed. International RILEM Conference on Material Science. Aachen, RILEM Publications S.A.R.L., 2010, 175–185.

[35] Snoeck D., Steuperaert S., Van Tittelboom K., Dubruel P., De Belie N. Visualization of water penetration in cementitious materials with superabsorbent polymers by means of neutron radiography. Cem Concr Res 2012, 42, 1113–1121.

[36] Snoeck D., Van den Heede P., Van Mullem T., De Belie N. Water penetration through cracks in self-healing cementitious materials with superabsorbent polymers studied by neutron radiography. Cem Concr Res 2018, 113, 86–98.

[37] Trtik P., Münch B., Weiss W.J., Kaestner A., Jerjen I., Josic L., Lehmann E., Lura P. Release of internal curing water from lightweight aggregates in cement paste investigated by neutron and X-ray tomography. Nucl Instrum Meth A 2011, 651, 244–249.

[38] Snoeck D., Van Tittelboom K., De Belie N., Steuperaert S., Dubruel P., The use of superabsorbent polymers as a crack sealing and crack healing mechanism in cementitious materials, in: Alexander M.G., Beushausen H.-D., Dehn F., Moyo P. Eds. 3rd International Conference on Concrete Repair, Rehabilitation and Retrofitting, Cape Town, 2012, pp. 152–157.

[39] Mönnig S. Water saturated super-absorbent polymers used in high strength concrete. Otto-Graf-J 2005, 16, 193–202.

[40] Mönnig S., Lura P. Superabsorbent polymers – An additive to increase the freeze-thaw resistance of high strength concrete. In: Grosse C.U., Ed. Adv Constr Mater. Berlin, Springer Berlin Heidelberg, 2007, 351–358.

[41] Mönnig S. Superabsorbing additions in concrete – applications, modelling and comparison of different internal water sources. In: Department of Civil Engineering. Stuttgart, The University of Stuttgart, 2009, 180.

[42] Snoeck D., Jensen O.M., De Belie N. The influence of superabsorbent polymers on the autogenous shrinkage properties of cement pastes with supplementary cementitious materials. Cem Concr Res 2015, 74, 59–67.

[43] Brüdern A.E., Mechtcherine V. Multifunctional use of SAP in strain-hardening cement-based composites. In: Jensen O.M., Hasholt M.T., Laustsen S., Eds. International RILEM Conference on Use of Superabsorbent Polymers and Other New Additives in Concrete. Lyngby, RILEM Publications S.A.R.L., 2010, 11–22.

[44] Mechtcherine V., Gorges M., Schröfl C., Assmann A., Brameshuber W., Bettencourt Ribeiro V., Cusson D., Custódio J., Fonseca da Silva E., Ichimiya K., Igarashi S., Klemm A., Kovler K., Lopes A., Lura P., Nguyen V.T., Reinhardt H.W., Filho R.D.T., Weiss J., Wyrzykowski M., Ye G., Zhutovsky S. Effect of internal curing by using Superabsorbent Polymers (SAP) on autogenous shrinkage and other properties of a high-performance fine-grained concrete: results of a RILEM Round-robin Test, TC 225-SAP. Mater Struct 2014, 47, 541–562.

[45] Snoeck D., Pel L., De Belie N. The water kinetics of superabsorbent polymers during cement hydration and internal curing visualized and studied by NMR. Sci Rep 2017, 7, 1–14.

[46] Lefever G., Snoeck D., De Belie N., Van Vlierberghe S., Van Hemelrijck D., Aggelis D.G. The contribution of elastic wave NDT to the characterization of modern cementitious media. Sensors 2020, 20, 2959.

[47]   Powers T.C., Brownyard T.L. Studies of the physical properties of hardened Portland cement paste, Portland Cement Association. Cornell, Research Laboratories, 1948.

[48]   De Meyst L., Mannekens E., Araújo M.A., Snoeck D., Van Tittelboom K., Van Vlierberghe S., Deroover G., De Belie N. Parameter study of superabsorbent polymers for use in durable concrete structures. Mater 2019, 12, 1541.

[49]   Tenório Filho J.R., Mannekens E., Van Tittelboom K., Snoeck D., De Belie N. Assessment of the potential of superabsorbent polymers as internal curing agents in concrete by means of optical fiber sensors. Constr Build Mater 2020, 238, 1–8.

[50]   Snoeck D., De Belie N. Effect of superabsorbent polymers, superplasticizer and additional water on the setting of cementitious materials. Int J of 3R's 2015, 5, 721–729.

[51]   Lura P., Ye G., Cnudde V., Jacobs P. Preliminary results about 3D distribution of superabsorbent polymers in mortars. In: Sun W., van Breugel K., Miao C., Ye G., Chen H., Eds. International Conference on Microstructure Related Durability of Cementitious Composites. Nanjing, 2008, 1341–1348.

[52]   Reinhardt H.W., Assmann A. Enhanced durability of concrete by superabsorbent polymers. In: Brandt A.M., Olek J., Marshal I.H., Eds. International Symposium Brittle Matrix Composites 9. Warsaw, Woodhead Publishing, 2009, 291–300.

[53]   Snoeck D., Schaubroeck D., Dubruel P., De Belie N. Effect of high amounts of superabsorbent polymers and additional water on the workability, microstructure and strength of mortars with a water-to-cement ratio of 0.50. Constr Build Mater 2014, 72, 148–157.

[54]   Snoeck D., Velasco L.F., Mignon A., Van Vlierberghe S., Dubruel P., Lodewyckx P., De Belie N. The influence of different drying techniques on the water sorption properties of cement-based materials. Cem Concr Res 2014, 64, 54–62.

[55]   Snoeck D., Velasco L.F., Mignon A., Van Vlierberghe S., Dubruel P., Lodewyckx P., De Belie N. The effects of superabsorbent polymers on the microstructure of cementitious materials studied by means of sorption experiments. Cem Concr Res 2015, 77, 26–35.

[56]   Romero Rodríguez C., Chaves Figueiredo S., Schlangen E., Snoeck D. Modeling water absorption in cement-based composites with SAP additions. In: Meschke G., Pickler B., Rots J.G., Eds. EURO-C 2018: computational Modelling of Concrete and Concrete Structures. Bad Hofgastein, 2018, 142, 141–110.

[57]   Romero Rodriguez C., Chaves Figueiredo S., Deprez M., Snoeck D., Schlangen E., Šavija B. Numerical investigation of crack self-sealing in cement-based composites with superabsorbent polymers. Cem Concr Comp 2019, 104, 1–12.

[58]   Snoeck D., Pel L., De Belie N. Comparison of different techniques to study the nanostructure and microstructure of cementitious materials with and without superabsorbent polymers. Constr Build Mater 2019, 223, 244–253.

[59]   Lura P., Durand F., Loukili A., Kovler K., Jensen O.M. Compressive strength of cement pastes and mortars with superabsorbent polymers. In: Jensen O.M., Lura P., Kovler K., Eds. International RILEM Conference on Volume Changes of Hardening Concrete: testing and Mitigation. Lyngby, RILEM Publications SARL, 2006, 117–125.

[60]   Snoeck D., De Schryver T., De Belie N. Enhanced impact energy absorption in self-healing strain-hardening cementitious materials with superabsorbent polymers. Constr Build Mater 2018, 191, 13–22.

[61]   Snoeck D., De Belie N. Repeated autogenous healing in strain-hardening cementitious composites by using superabsorbent polymers. J Mater Civ Eng 2015, 04015086, 1–11.

[62]   Mechtcherine V. Use of superabsorbent polymers (SAP) as concrete additive. RILEM Tech Lett 2016, 1, 81–87.

[63]   He Z., Shen A., Guo Y., Lyu Z., Li D., Qin X., Zhao M., Wang Z. Cement-based materials modified with superabsorbent polymers: a review. Constr Build Mater 2019, 225, 569–590.

[64] Lura P., Jensen O.M., Igarashi S. Experimental observation of internal curing of concrete. Mater Struct 2007, 40, 211–220.

[65] Bentz D.P., Weiss W.J. Internal Curing: A 2010 State-of-the-Art Review. In: Interagency N., Ed. 2011, 82.

[66] Snoeck D. Superabsorbent polymers to seal and heal cracks in cementitious materials. RILEM Tech Lett 2018, 3, 32–38.

[67] Snoeck D., Van Tittelboom K., Steuperaert S., Dubruel P., De Belie N. Self-healing cementitious materials by the combination of microfibres and superabsorbent polymers. J Intel Mat Syst Str 2014, 25, 13–24.

[68] Farzanian K., Pimenta Teixeira K., Perdigão Rocha I., De Sa Carneiro L., Ghahremaninezhad A. The mechanical strength, degree of hydration, and electrical resistivity of cement pastes modified with superabsorbent polymers. Constr Build Mater 2016, 109, 156–165.

[69] AzariJafari H., Kazemian A., Rahimi M., Yahia A. Effects of pre-soaked super absorbent polymers on fresh and hardened properties of self-consolidating lightweight concrete. Constr Build Mater 2016, 113, 215–220.

[70] Kong X.-M., Zhang Z.-L., Lu Z.-C. Effect of pre-soaked superabsorbent polymer on shrinkage of high-strength concrete. Mater Struct 2015, 48, 2741–2758.

[71] Ma X., Yuan Q., Liu J., Shi C. Effects of SAP on the properties and pore structure of high performance cement-based materials. Constr Build Mater 2017, 131, 476–484.

[72] Oh S., Choi Y.C. Superabsorbent polymers as internal curing agents in alkali activated slag mortars. Constr Build Mater 2018, 159, 1–8.

[73] Wehbe Y., Ghahremaninezhad A. Combined effect of shrinkage reducing admixtures (SRA) and superabsorbent polymers (SAP) on the autogenous shrinkage, hydration and properties of cementitious materials. Constr Build Mater 2017, 138, 151–162.

[74] Almeida F.D.C.R., Klemm A.J., Effect of superabsorbent polymers (SAP) on fresh state mortars with ground granulated blast-furnace slag (GGBS), in: 5th International Conference on Durability of Concrete Structures, Shenzhen, 2016, pp. 1–7.

[75] Almeida F.C.R., Klemm A. Efficiency of internal curing by superabsorbent polymers (SAP) in PC-GGBS mortars. Cem Concr Comp 2018, 88, 41–51.

[76] Kang S.-H., Hong S.-G., Moon J. The effect of superabsorbent polymer on various scale of pore structure in ultra-high performance concrete. Constr Build Mater 2018, 172, 29–40.

[77] Yao Y., Zhu Y., Yang Y. Incorporation of SAP particles as controlling pre-existing flaws to improve the performance of ECC. Constr Build Mater 2011, 28, 139–145.

[78] Gruyaert E., Debbaut B., Snoeck D., Díaz P., Arizo A., Tziviloglou E., Schlangen E., De Belie N. Self-healing mortar with pH-sensitive superabsorbent polymers: testing of the sealing efficiency of self-healing mortars by water flow tests. Smart Mater Struct 2016, 25(084007), 084001–084011.

[79] Mignon A., Snoeck D., Schaubroeck D., Luickx N., Dubruel P., Van Vlierberghe S., De Belie N. pH-responsive superabsorbent polymers: a pathway to self-healing of mortar. React Funct Polym 2015, 93, 68–76.

[80] Mignon A., Graulus G.-J., Snoeck D., Martins J., De Belie N., Dubruel P., Van Vlierberghe S. pH-sensitive superabsorbent polymers: a potential candidate material for self-healing concrete. J Mater Sci 2015, 50, 970–979.

[81] Mignon A., Snoeck D., D'Halluin K., Balcaen L., Vanhaecke F., Dubruel P., Van Vlierberghe S., De Belie N. Alginate biopolymers: counteracting the impact of superabsorbent polymers on mortar strength. Constr Build Mater 2016, 110, 169–174.

[82] Mignon A., Vermeulen J., Snoeck D., Dubruel P., Van Vlierberghe S., De Belie N. Mechanical and self-healing properties of cementitious materials with pH-responsive semi-synthetic superabsorbent polymers. Mater Struct 2017, 50, 1–12.

[83] Pelto J., Leivo M., Gruyaert E., Debbaut B., Snoeck D., De Belie N. Application of encapsulated superabsorbent polymers in cementitious materials for stimulated autogenous healing Smart. Mater Struct 2017, 26, 1–14.

[84] Liu H., Bu Y., Sanjayan J.G., Nazari A., Shen Z. The application of coated superabsorbent polymer in well cement for plugging the microcrack. Constr Build Mater 2016, 104, 72–84.

[85] Hu M., Guo J., Du J., Liu Z., Li P., Ren X., Feng Y. Development of Ca2+-based, ion-responsive superabsorbent hydrogel for cement applications: self-healing and compressive strength. J Colloid Interface Sci 2019, 538, 397–403.

[86] Pourjavadi A., Fakoorpoor S.M., Hosseini P., Khaloo A. Interactions between superabsorbent polymers and cement-based composites incorporating colloidal silica nanoparticles. Cem Concr Comp 2013, 37, 196–204.

[87] Lefever G., Snoeck D., Aggelis D.G., De Belie N., Van Vlierberghe S., Van Hemelrijck D. Evaluation of the self-healing ability of mortar mixtures containing superabsorbent polymers and nanosilica. Mater 2020, 13, 380.

[88] Lefever G., Tsangouri E., Snoeck D., Aggelis D., De Belie N., Van Vlierberghe S., Van Hemelrijck D. Combined use of superabsorbent polymers and nanosilica for reduction of restrained shrinkage and strength compensation in cementitious mortars. Constr Build Mater 2020, 251, 118966.

[89] Mechtcherine V., Schröfl C., Wyrzykowski M., Gorges M., Cusson D., Margeson J., De Belie N., Snoeck D., Ichimiya K., Igarashi S.-I., Falikman V., Friedrich S., Bokern J., Kara P., Lura P., Marciniak A., Reinhardt H.-W., Sippel S., Ribeiro A.B., Custódio J., Ye G., Dong H., Weiss J. Effect of superabsorbent polymers (SAP) on the freeze-thaw resistance of concrete: results of a RILEM interlaboratory test. Mater Struct 2017, 50, 1–19.

[90] Laustsen S., Hasholt M.T., Jensen O.M. Void structure of concrete with superabsorbent polymers and its relation to frost resistance of concrete. Mater Struct 2015, 48, 357–368.

[91] Wyrzykowski M., Terrasi G., Lura P. Expansive high-performance concrete for chemical-prestress applications. Cem Concr Res 2018, 107, 275–283.

[92] Craeye B., Tielemans T., Lauwereijssens G., Stoop J. Effect of super absorbing polymers on the freeze-thaw resistance of coloured concrete roads. Road Materials and Pavement Design 2013, 14, 90–106.

[93] Li F., Liu J., Yuan J., He X., Zhang R. Durability of concrete modified with superabsorbent polymers. Mater Perform 2018, 57, 54–58.

[94] Craeye B., Cockaerts G., Kara De Maeijer P. Improving freeze–thaw resistance of concrete road infrastructure by means of superabsorbent polymers. Infrastructures 2018, 3, 4.

[95] Kusayama S., Kuwabara H., Igarashi S.-I., Comparison of salt scaling resistance of concretes with different types of superabsorbent polymers, in: Mechtcherine V., Schröfl C. (Eds.) Application of Superabsorbent Polymers and Other New Admixtures in Concrete Construction, RILEM Publications S.A.R.L., 2014.

[96] Tan Y., Chen H., Wang Z., Xue C., He R. Performances of cement mortar incorporating Superabsorbent Polymer (SAP) using diferent dosing methods. Mater 2019, 12, 1619.

[97] Tenório Filho J.R., Snoeck D., De Belie N. Mixing protocols for plant-scale production of concrete with superabsorbent polymers. Struct Concr 2020, 21, 983–991.

[98] Lee H.X.D., Wong H.S., Buenfeld N.R. Potential of superabsorbent polymer for self-sealing cracks in concrete. Adv Appl Ceram 2010, 109, 296–302.

[99] Lee H.X.D., Wong H.S., Buenfeld N.R. Self-sealing cement-based materials using superabsorbent polymers. In: Jensen O.M., Hasholt M.T., Laustsen S., Eds. International RILEM Conference on Use of Superabsorbent Polymers and Other New Additives in Concrete. Lyngby, RILEM Publications SARL, 2010, 171–178.

[100] Lee H.X.D., Wong H.S., Buenfeld N.R. Self-sealing of cracks in concrete using superabsorbent polymers. Cem Concr Res 2016, 79, 194–208.
[101] Snoeck D., Dubruel P., De Belie N. Superabsorbent polymers to prevent water movement in cementitious materials. Int J of 3R's 2012, 3, 432–440.
[102] Al-Nasra M., Daoud M. Study of the ability of cracked concrete to block water flow, concrete mixed with super absorbent polymer. ARPN Int J Eng Appl Sci 2017, 12, 274–281.
[103] Hong G., Choi S. Rapid self-sealing of cracks in cementitious materials incorporating superabsorbent polymers. Constr Build Mater 2017, 143, 366–375.
[104] Hong G., Song C., Park J., Choi S. Hysteric behavior of rapid self-sealing of cracks in cementitious materials incorporating superabsorbent polymers. Constr Build Mater 2019, 195, 187–197.
[105] Snoeck D., Alderete N.M., Van Belleghem B., Van den Heede P., Van Tittelboom K., De Belie N., Internal curing of cement pastes by superabsorbent polymers studied by means of neutron radiography in: 14th International Conference on Durability of Building Materials and Components, Gent, 2017, pp. 1–8.
[106] Song X.F., Wei J.F., He T.S. A method to repair concrete leakage through cracks by synthesizing super-absorbent resin in situ. Constr Build Mater 2009, 23, 386–391.
[107] Van Tittelboom K., Wang J., Gomes M.A., Araújo D., Snoeck D., Gruyaert E., Debbaut B., Derluyn H., Cnudde V., Tsangouri E., Hemelrijck D.V., Belie N.D. Comparison of different approaches for self-healing concrete in a large-scale lab test. Constr Build Mater 2016, 108, 125–137.
[108] Hearn N. Self-sealing, autogenous healing and continued hydration: what is the difference? Mater Struct 1998, 31, 563–567.
[109] Yang E.-H. Designing added functions in engineered cementitious composites. In: Civil Engineering. Ann Arbor, University of Michigan, 2008, 293.
[110] Ter Heide N. Crack healing in hydrating concrete. In: Civil Engineering and Geosciences, Microlab. Delft, Delft University of Technology, 2005, 128.
[111] Homma D., Mihashi H., Nishiwaki T. Self-healing capability of fibre reinforced cementitious composites. J Adv Concr Technol 2009, 7, 217–228.
[112] Edvardsen C. Water permeability and autogenous healing of cracks in concrete. ACI Mater J 1999, 96, 448–454.
[113] Granger S., Loukili A., Pijaudier-Cabot G., Chanvillard G. Experimental characterization of the self-healing of cracks in an ultra high performance cementitious material: mechanical tests and acoustic emission analysis. Cem Concr Res 2007, 37, 519–527.
[114] Lauer K.R., Slate F.O. Autogenous healing of cement paste. ACI Mater J 1956, 52, 1083–1097.
[115] Neville A. Autogenous healing – A concrete miracle? Concr Int 2002, 24, 76–82.
[116] van Breugel K., Is there a market for self-healing cement-based materials? in: 1st International Conference on self-healing Materials, RILEM Publications SARL, Noordwijk, 2007, pp. 1–9.
[117] Jia H.Y., Wei C., Ming X.Y., Yang E.-H. The Microstructure of Self-Healed PVA ECC Under Wet and Dry Cycles. Mater Res 2010, 13, 225–231.
[118] Snoeck D., De Belie N. Autogenous healing in strain-hardening cementitious materials with and without superabsorbent polymers: an 8-year study. Front Mater 2019, 6, 1–12.
[119] Jooss M. Leaching of concrete under thermal influence. Otto-Graf-J 2001, 12, 51–68.
[120] Snoeck D., Criel P. Voronoi diagrams and self-healing cementitious materials: a perfect match. Adv Cem Res 2018, 31, 261–269.
[121] Jacobsen S., Marchand J., Hornain H. SEM observations of the microstructure of frost deteriorated and self-healed concrete. Cem Concr Res 1995, 25, 55–62.

[122] Jacobsen S., Sellevold E.J. Self-healing of high-strength concrete after deterioration by freeze/thaw. Cem Concr Res 1996, 26, 55–62.

[123] Yang Y., Lepech M.D., Yang E.-H., Li V.C. Autogenous healing of engineered cementitious composites under Wet-dry cycles. Cem Concr Res 2009, 39, 382–390.

[124] Yang Y., Yang E.-H., Li V.C. Autogenous healing of engineered cementitious composites at early age. Cem Concr Res 2011, 41, 176–183.

[125] Kim J.S., Schlangen E. Super absorbent polymers to simulate self healing in ECC. In: van Breugel K., Ye G., Yuan Y., Eds. 2nd International Symposium on Service Life Design for Infrastructures. Delft, RILEM Publications SARL, 2010, 849–858.

[126] Termkhajornkit P., Nawa T., Yamashiro Y., Saito T. Self-healing ability of fly ash-cement systems. Cem Concr Comp 2009, 31, 195–203.

[127] Van den Heede P., Maes M., De Belie N. Influence of active crack width control on the chloride penetration resistance and global warming potential of concrete slabs made with fly ash + silica fume concrete. Constr Build Mater 2014, 67, 74–80.

[128] Van Tittelboom K., Gruyaert E., Rahier H., De Belie N. Influence of mix composition on the extent of autogenous crack healing by continued hydration or calcium carbonate formation. Constr Build Mater 2012, 37, 349–359.

[129] Yildirim G., Sahmaran M., Ahmed H. Influence of hydrated lime addition on the self-healing capability of high-volume fly ash incorporated cementitious composites. J Mater Civ Eng 2014, 04014187, 1–11.

[130] Gruyaert E., Tittelboom K., Rahier H., Belie N. Activation of the pozzolanic or latent-hydraulic reaction by alkalis in order to repair cracks in concrete. J Mater Civ Eng 2014, 04014208.

[131] Ahn T.-H., Kishi T. Crack self-healing behavior of cementitious composites incorporating various mineral admixtures. J Adv Concr Technol 2010, 8, 171–186.

[132] Ferrara L., Krelani V., Carsana M. A "fracture testing" based approach to assess crack healing of concrete with and without crystalline admixtures. Constr Build Mater 2014, 68, 535–551.

[133] Krelani V. Self-healing capacity of cementitious composites. In: Department of Civil & Environmental Engineering. Milano, Politecnico Di Milano, 2015, 256.

[134] Ferrara L. Crystalline admixtures in cementitious composites: from porosity reducers to catalysts of self healing. In: Mechtcherine V., Schröfl C., Eds. International RILEM Conference on the Application of Superabsorbent Polymers and Other New Admixtures in Concrete Construction. Dresden, RILEM Publications S.A.R.L., 2014, 311–324.

[135] Roig-Flores M., Moscato S., Serna P., Ferrara L. Self-healing capability of concrete with crystalline admixtures in different environments. Constr Build Mater 2015, 86, 1–11.

[136] Sisomphon K., Çopuroğlu O., Koenders E.A.B. Self-healing of surface cracks in mortars with expansive additive and crystalline additive. Cem Concr Comp 2012, 34, 566–574.

[137] Gwon S., Ahn E., Shin M. Self-healing of modified sulfur composites with calcium sulfoaluminate cement and superabsorbent polymer. Composites Part B 2019, 162, 469–483.

[138] Wang F., Bu Y., Guo S., Lu Y., Sun B., Shen Z. Self-healing cement composite: amine- and ammonium-based pH-sensitive superabsorbent polymers. Cem Concr Comp 2019, 96, 154–165.

[139] Wang J., Mignon A., Snoeck D., Wiktor V., Boon N., De Belie N. Application of modified-alginate encapsulated carbonate producing bacteria in concrete: a promising strategy for crack self-healing. Front Microbiol 2015, 6, 1–14.

[140] Wang J., Snoeck D., Van Vlierberghe S., Verstraete W., De Belie N. Application of hydrogel encapsulated carbonate precipitating bacteria for approaching a realistic self-healing in concrete. Constr Build Mater 2014, 68, 110–119.

[141] Li V.C., Lim Y.M., Chan Y.W. Feasibility study of a passive smart self-healing cementitious composite. Composites Part B 1988, 29, 819–827.

[142] Li V.C., Leung C.K.Y. Steady state and multiple cracking of short random fiber composites. ASCE Eng Mech 1992, 188, 2246–2264.
[143] Li V.C., Wang S., Wu C. Tensile strain-hardening behavior of polyvinyl alcohol engineered cementitious composites (PVA-ECC). ACI Mater J 1997, 98, 483–492.
[144] Li V.C., Horii H., Kabele P., Kanda T., Lim Y.M. Repair and retrofit with engineered cementitious composites. Eng Fract Mech 2000, 65, 317–334.
[145] Li V.C., Wu C., Wang S., Ogawa A., Saito T. Interface Tailoring for Strain-Hardening Polyvinyl Alcohol – Engineered Cementitious Composites (PVA-ECC). ACI Mater J 2002, 99, 463–472.
[146] Li V.C. Engineered Cementitious Composites (ECC) – material, structural, and durability performance. In: Nawy E., Ed. Concrete Construction Engineering Handbook. CRC Press, 2008, 78.
[147] Li V.C., Herbert E. Robust self-healing concrete for sustainable infrastructure. J Adv Concr Technol 2012, 10, 207–218.
[148] Luković M., Dong H., Šavija B., Schlangen E., Ye G., van Breugel K. Tailoring strain-hardening cementitious composite repair systems through numerical experimentation. Cem Concr Comp 2014, 53, 200–213.
[149] Snoeck D., De Belie N. Mechanical and self-healing properties of cementitious composites reinforced with flax and cottonised flax, and compared with polyvinyl alcohol fibres. Biosyst Eng 2012, 111, 325–335.
[150] Snoeck D., Smetryns P.-A., De Belie N. Improved multiple cracking and autogenous healing in cementitious materials by means of chemically-treated natural fibres. Biosyst Eng 2015, 139, 87–99.
[151] Snoeck D., Debo J., De Belie N. Translucent self-healing cementitious materials using glass fibers and superabsorbent polymers. Dev Built Environ 2020, 3, 1–8.
[152] Snoeck D. Self healing concrete by the combination of microfibres and reactive substances. In: Engineering and Architecture, Structural Engineering. Ghent, Ghent University, 2011, 199.
[153] Deng H., Liao G. Assessment of influence of self-healing behavior on water permeability and mechanical performance of ECC incorporating superabsorbent polymer (SAP) particles. Constr Build Mater 2018, 170, 455–465.
[154] Snoeck D., De Belie N. Effect of fibre type and superabsorbent polymers on the self-healing properties of strain-hardening cementitious materials. In: Brandt A.M., Olek J., Glinicki M.A., Leung C.K.Y., Lis J., Eds. Proceedings of the International Symposium on Brittle Matrix Composites, BMC11. Warsaw, 2015, 213–222.
[155] Snoeck D., Dewanckele J., Cnudde V., De Belie N. X-ray computed microtomography to study autogenous healing of cementitious materials promoted by superabsorbent polymers. Cem Concr Comp 2016, 65, 83–93.
[156] Snoeck D., Pel L., De Belie N. Autogenous healing in cementitious materials with superabsorbent polymers quantified by means of NMR. Sci Rep 2020, 10, 1–6.
[157] Mechtcherine V., Dudziak L., Schulze J., Stärhr H. Internal curing by Super Absorbent Polymers – Effects on material properties of self-compacting fibre-reinforced high performance concrete. In: Jensen O.M., Lura P., Kovler K., Eds. international RILEM conference on Volume Changes of Hardening Concrete: Testing and Mitigation. Lyngby, RILEM publications S.A.R.L., 2006, 87–96.
[158] Zhu C., Li X., Xie Y. Influence of SAP on the performance of concrete and its application in Chinese railway construction. In: Mechtcherine V., Schröfl C., Eds. Application of Superabsorbent Polymers and Other New Admixtures in Concrete Construction. Dresden, RILEM Publications S.A.R.L., 2014, 345–354.
[159] Liu R., Sun Z., Ding Q., Chen P., Chen K. Mitigation of early-age cracking of concrete based on a new gel-type superabsorbent polymer. J Mater Civ Eng 2017, 29, 04017151.

[160] Liu J., Yu C., Shu X., Ran Q., Yang Y., Recent advance of chemical admixtures in concrete, in: Gemrich J. (Ed.) 15th International Congress on the Chemistry of Cement, Research Institute of Binding Materials Prague, Prague, 2019, pp. 12.
[161] Van den Heede P., Mignon A., Habert G., De Belie N. Cradle-to-gate life cycle assessment of self-healing engineered cementitious composite with in-house developed (semi-)synthetic superabsorbent polymers. Cem Concr Comp 2018, 94, 166–180.
[162] Altmann F., Mechtcherine V. Durability design strategies for new cementitious materials. Cem Concr Res 2013, 54, 114–125.

Index

https://doi.org/10.1515/9781501519116-006

www.ingramcontent.com/pod-product-compliance
Lightning Source LLC
Chambersburg PA
CBHW081530220326
41598CB00036B/6388